# LAND DEGRADATION IN THE DINDER AND RAHAD BASINS: INTERACTIONS BETWEEN HYDROLOGY, MORPHOLOGY AND ECOHYDROLOGY IN THE DINDER NATIONAL PARK, SUDAN



KHALID ELNOOR ALI HASSABALLAH

LAND DEGRADATION IN THE DINDER AND RAHAD BASINS:
INTERACTIONS BETWEEN HYDROLOGY, MORPHOLOGY AND
ECOHYDROLOGY IN THE DINDER NATIONAL PARK, SUDAN

DISSERTATION

Submitted in fulfillment of the requirements of
the Board for Doctorates of Delft University of Technology
and
of the Academic Board of the IHE Delft
Institute for Water Education
for
the Degree of DOCTOR
to be defended in public on
Monday, 09 November 2020, at 15:00 hours
in Delft, the Netherlands

by

Khalid Elnour Ali HASSABALLAH
Master of Science Degree in Hydroinformatics, UNESCO-IHE Institute for Water Education, Delft, the Netherlands
born in Sennar, Sudan

This dissertation has been approved by the
promotor: Prof. dr. S. Uhlenbrook and
copromotor: Dr. Y.A. Mohamed

Composition of the doctoral committee:

| | |
|---|---|
| Rector Magnificus TU Delft | Chairman |
| Rector IHE Delft | Vice-Chairman |
| Prof. dr. S. Uhlenbrook | TU Delft /IHE Delft, promotor |
| Dr. Y.A. Mohamed | IHE Delft, copromotor |

Independent members:

| | |
|---|---|
| Prof.dr. S. Hamad Abdalla | U-Khartoum |
| Prof.dr. M.J. Franca | TU Delft /IHE Delft |
| Prof.dr.ir. N.C. van de Giesen | TU Delft |
| Prof.dr.ir. P. van der Zaag | TU Delft /IHE Delft |
| Prof. dr. D. P. Solomatine | TU Delft /IHE Delft, reserve member |

*This research was conducted under the auspices of the Graduate School for Socio-Economic and Natural Sciences of the Environment (SENSE)*

CRC Press/Balkema is an imprint of the Taylor & Francis Group, an informa business

Published by:
CRC Press/Balkema
Schipholweg 107C, 2316 XC, Leiden, the Netherlands
Pub.NL@taylorandfrancis.com
www.crcpress.com – www.taylorandfrancis.com

ISBN: 978-0-367-68355-9

***Dedicated to the loving memory of my father. You are not here to cheer me, but I made it - just like you said I would.***

# ACKNOWLEDGMENTS

First of all, I would like to thank Allah (God) for His never-ending care and giving me stamina to accomplish this research well in a given period of time. This research would not have been possible without proper guidance, support, and encouragements from different people and organizations. My heartfelt gratitude to my promoter Prof. dr. Stefan Uhlenbrook, who has given me a chance to do my PhD under his guidance. Stefan, I appreciate your intellectual scientific capabilities and the response to all the academic and non-academic matters related to my research. I have learned a lot from you. You are always optimistic and see things in different angles. Your expertise knowledge as a hydrologist helped me a lot to understand how catchment hydrological studies work and particularly helped me to better understand data scarce environments.

I am also indebted to my supervisor Prof. dr. Yasir A. Mohamed who has been appointed as a Minister of Irrigation and Water Resources, Sudan during the final stage of my PhD, for his valuable contribution to my research in many ways. I well-regarded his broad knowledge related to water management. I really appreciate his consistent guidance, critical comments and suggestions to bring the dissertation to the current form. Dr. Yasir, I appreciated your knowledge of basin hydrology in general and the Blue Nile in particular, that helped me a lot to interpret the results in a more scientific way.

I am also thankful to the Netherlands Fellowship Program (NFP), for financing most of the costs related to this research throughout the research period. I would take this opportunity to thank Ms. Jolanda Boots, PhD fellowship officer at IHE-Delft, who managed the administrative and financial issues related to my research work.

I would like to thank my employer the Hydraulics Research Center of the Ministry of Irrigation and Water Resources, Sudan, for granting me a study leave to execute this research work. My sincere thanks also go to all the PhD researchers, namely Dr. Yasir Salih, Dr. Sirak Tekleab, Dr. Abonesh Tesfaye, Dr. Ermias Teferi, Dr. Hermen Smit, Dr. Eshraga Sokarab, Dr. Reem Degna and Rahel Haile who have been conducting their research in the Blue Nile hydro-solidarity project. I really appreciate their cooperation and the efforts we have made jointly to generate and share multi-disciplinary knowledge about the Blue Nile Basin. I extend my special thanks to all PhD research fellows and MSc participants at IHE-Delft. I am also grateful to Dr. Hermen Smit for translating the summary into the Dutch language.

Special thanks go to my colleagues at the Hydraulics Research Center (HRC) for their support during all stages of this research. Special thanks to Mr. Khalid Abdelwahab and the late technician Mr. Khamis Adam and other members of the technician team at HRC for joining me during the fieldwork in a very remote environment in the Dinder National Park in Sudan, Mr. Adil Dawoud for analyzing the sediment samples that I have collected from the Dinder and Rahad. The fieldwork activities would not have been possible without the support I got from the Wildlife Conservation General Administration (WCGA), Sudan, specially from Mr. Jamal El Balla. I am also thankful to Dr. Omer Meina from the Wildlife Research Center who helped me in the field survey for collecting the flora's data.

I have been very lucky to have many friends who continuously supported me and shared pleasant times with me. Thank you, my friends: Dr. Chol Abel, Eng. Amgad Omer, Eng. Safwan Elsagier, Eng. Nazar Naiema, Dr. Tesfay Gebremicael, Dr. Eiman Fadoul, Eng. Shaza Gameel, Eng. Salman Fadlelmoula, and Eng. Sami Osman. I wish to extent my gratitude to you, for always supporting me and for creating a very convenient environment, and for always being decent and wise friends. Thanks also to any other friends who I haven't mentioned, but who have supported and encouraged me.

Last but not least, I would like to thank my wife for her consistent love, care, encouragements and patience throughout the research period. My lovely daughters *Aya, Shahd and Ruba* who missed me a lot and I missed them as well while I was in the Netherlands.

Finally, I would like to express my deepest gratefulness and appreciation to my extended family; my lovely mother, brothers and sisters for their endless source of love, encouragement, inspiration, patience and deep understanding during all these years.

Khalid Elnoor Ali Hassaballah

June, 2020, Wad Medani, Sudan

# SUMMARY

The headwater catchments of the Dinder and Rahad river basins (D&R) generate over 7% of the Blue Nile water. The two basins are shared between Ethiopia and Sudan and cover and area of about 77,504 km$^2$. The Rahad river provides water to the Rahad irrigation scheme in Sudan (126,000 ha), while the Dinder river supports the diverse ecosystem of the Dinder National Park (DNP) covering an area of about 10,291 km$^2$. However, the two rivers experienced significant changes of the floodplain hydrology during recent years. This has large implications on the ecosystem of the so-called "*mayas*" located in the DNP. Maya is a local name for floodplain wetlands and oxbows cut off from the meandering river that are found on both sides of the Dinder river and its tributaries. Mayas are important ecosystems in the park as they constitute the main source of food and water for wildlife during the dry season which extends from November to June. Appropriate water resources development and sustainable ecosystem conservation should consider the climate variability and the land use and land cover (LULC) changes and their impacts on catchment response. Unlike for the Blue Nile river, very few studies have been carried out for the two catchments of Dinder and Rahad.

The Dinder and Rahad river basins have a complex hydrology, with varying climate, topography, soil, vegetation and geology as well as human interventions. Although the area is blessed with a variety of natural resources, it is facing damaging human activities such as intensive grazing, deforestation, and improper farming practices on the steep slopes. These human practices have posed a great threat to the sustainability and the ecosystem integrity and subsequently influenced the wildlife and plant species in the diverse ecosystem of the DNP.

The spatial and temporal variability of the hydro-climate as well as land use changes are among the most challenging problems. Understanding the interaction between climate, LULC changes and their links to hydrology, river morphology and ecohydrology in the Dinder and Rahad basins is difficult given the lack of climatic, hydrological and ecological data. The hydrological processes of the basin are not fully understood, in particular the prediction of hydrological dynamics under current conditions as well as under future changes. Therefore, in-depth hydrological studies of the basins are crucial for planning and management of water resources as well as the environment.

This research investigated the impacts of land degradation on the Dinder and Rahad hydrology and morphology, and interlinkage to the ecohydrological system of the DNP-Sudan. It used an ensemble of techniques to improve our understanding of the hydrological processes and LULC changes in the basins. This included long-term trend analysis of hydroclimatic variables, land use and land cover changes analysis, field measurements, rainfall-runoff modelling, GIS and remote sensing data acquisition and analysis, hydrodynamic and morphological modelling of the Dinder river and its floodplain, with special focus on the maya wetlands. Moreover, this research is the first study to investigate the eco-hydrology of the DNP. It is expected that the output of the study will be beneficial to all stakeholders concerned and support decision-making processes for better understanding and management of water resources and ecosystem conservation in the area and possibly beyond.

The long-term trends of the hydro-climatology of the Dinder and Rahad basins were assessed. The non-parametric Mann-Kendall (MK) and Pettitt tests were applied to analyze the trends and the change points of hydro-climatic data time series of streamflow, rainfall and temperature. Trends have been assessed at 5% significance level for different time periods and varying lengths based on data availability. The indicators of hydrologic alterations (IHA) approach (Richter et al., 1996) was applied to verify the MK test and to analyze the essential characteristics of the streamflow likely to impact ecological functions in the D&R basins, including flow magnitude, flow timing and rate of change in river flows. Understanding the level to which the streamflow has changed from its natural conditions is crucial for developing an effective management plan for the ecosystem conservation/restoration.

Streamflow of the Rahad river exhibited statistically significant increasing trend for the period 1972-2011, while no evidence for significant trend in the Dinder river. Nevertheless, the analysis of monthly maxima showed a shift towards decreased flows during the high flow period and increased flows during the low flow period. The Dinder maxima during the high flow period (August flow) decreased from 517 $m^3/s$ during the early part of the record (1972-1991) to 396 $m^3/s$ during the latest years (1992-2011). Temperature showed significantly increasing trends at the rate of 0.24 and 0.30 °C/decade for the two examined stations. Rainfall showed no significant change.

The IHA-based analysis has shown that the flow of the Rahad river was associated with significant upward alterations in some of the hydrological indicators. The flow of the Dinder river was associated with significant downward alterations. Particularly, these were: a) a decrease in the magnitude of the river flow during the high flow period (August flow) and an increase in low flows (November flow); b) a decrease in magnitude of flow extremes (i.e. 1, 7, 30 and 90-day maxima); and c) a decrease in flow rise rate and an increase in flow fall rate. These alterations in the Dinder river flows are likely to affect the ecosystems in DNP negatively. The trend analysis results suggest other factors than climate variability (e.g. land use and land cover changes) to be responsible for streamflow alterations.

To understand the LULC changes and their consequence on the surface hydrology of the Dinder and Rahad basins, analysis of streamflow response to land use and land cover changes using satellite data and hydrological modelling was performed. The WFlow hydrological model was calibrated and run with different land use and land cover maps from 1972, 1986, 1998 and 2011 with fixed model parameters. Catchment topography, soil and land cover maps were derived from satellite images and served to estimate model parameters. Results of the LULC change detection between 1972 and 2011 indicate a significant increase in cropland and decrease in woodland. Cropland increased from 14% to 47% and from 18% to 68% in Dinder and Rahad, respectively. Woodland decreased from 42% to 14% and from 35% to 14% for Dinder and Rahad, respectively. The model results show that streamflow is affected by LULC changes in both the Dinder and the Rahad rivers from decreasing to increasing flow according to LULC changes. The LULC changes significantly increased the streamflow during the years 1986 and 2011, mainly in the Rahad river. This could be attributed to the large decrease in woodland from 35% in 1972 to 14% in 1986, and the large expansion in cropland in the Rahad catchment to 68% of the total area in 2011, particularly in rain-fed cropland.

In the Dinder river, the annual streamflow increased by 20% between 1972 and 1986 but is followed by a decrease of 9% between 1986 and 1998. The increase in the annual streamflow between 1972 and 1986 could be a result of an increase in cropland, grassland and shrub land by 6%, 10%, 83%, respectively, associated with a decrease in woodland by 43% from 42% in 1972 to 23% in 1986. Over the period 1986–1998, woodland and cropland increased by 16% and 192%, respectively, while the remaining land cover categories showed declines. Over the period 1998–2011, the annual streamflow increased by 52% and corresponds with findings on increases in the percentage of cropland, shrub land and bare land by 4%, 71% and 360%, respectively, while a decrease in grassland and woodland by 76% and 50%, respectively. The decrease in percentage change of bare area over the period 1986–1998, along with the increase in woodland in both the Dinder and the Rahad basins, indicates that the environment was recovering from the severe drought of 1984–1985.

A quasi 3D model was used to understand the morphological changes and hence support decision-making for the management of the maya ecosystems. In particular, the effect of morphological changes on both the Dinder river and the maya wetlands. The model extent covered an area of about 105 $km^2$ inside the DNP. SRTM, digital elevation model (DEM 90 m) was used along a 20-km reach of the Dinder river, Sudan. Since the vertical accuracy of the 90 m DEM performs poorly in areas of moderate topographic variation and forested area, two field topographic surveys were conducted during the years 2013 and 2016 using levelling and geographic information system (GPS) devices to generate a DEM with higher accuracy (vertical error of 0.008 m and horizontal error of ± 3 m) for the model domain within the DNP. The intersection of these data with a high vertical accuracy survey of floodplain topography obtained through the field surveys permitted the simulation of the maya wetlands filling and emptying mechanism.

Due to the absence of water level measurements within the pilot area inside the DNP, a monitoring network utilizing so-called Divers was established in June 2013 to measure water level data for this research. The network consists of two Mini-Divers for recording water level, temperature and atmospheric pressure measurements and one Baro-Diver to measure the atmospheric pressure that is used to compensate for the variations in atmospheric pressure. The recorded water levels were used to calibrate the hydrodynamic model.

To understand the hydrological and morphological connectivity of the maya in terms of filling/emptying and sediment transport processes, six scenarios were analyzed. The first three scenarios consider three different hydrologic conditions of wet, average and dry years for the existing system with the constructed connection canal. The other three scenarios consider the same hydrologic conditions but for the natural system without the connection canal. The comparison between scenarios demonstrated that the hydrodynamics and sedimentology of the maya are driven by two factors: a) the hydrological variability of the Dinder river; and b) deposited sediment at the inlet channel of the natural drainage network.

Finally, the ecohydrology of maya wetlands in the DNP was assessed and relations between vegetation dynamics, wildlife and water availability were identified. To assess the ecosystem status and patterns of change, field data on vegetation composition and wildlife were collected from four mayas namely; Ras Amir maya, Musa maya, Gererrisa maya, and Abdelghani maya. To determine the status of functioning of the mayas, a systematic-random quadrat (SRQ) method was used to collect flora's data (indicators) from four mayas inside the DNP. The normalized difference water index (NDWI) was used to estimate the inundation extent and the normalized difference vegetation index (NDVI) was used to estimate the related vegetation coverage in the pilot Musa maya. Data on wildlife censuses in the four mayas were analyzed and relations to hydrological variability and vegetation cover were identified. The SRQ survey distinguished seven plant species in the four surveyed mayas, with floristic composition of plant species that considerably varies across the studied mayas. The NDVI analysis of the data between 2001 and 2016 showed significant variations in the area of vegetation cover. These variations were strongly linked to variations in the NDWI. The wildlife censuses showed that the population size and distribution of wildlife in the DNP depend mainly on the availability of water and pasture which are affected by hydrological variability. 84% of the total wildlife (herbivores) populations were found in the grassland within the periphery of mayas compared to only 16% in the burnt and open areas. This indicates that herbivores prefer grassland and woodland around the mayas rather than burnt and open areas. This is likely due to the availability of water, food (pasture) and shelter. Therefore, hydrological variability seems to be a key factor controlling the ecological processes.

Given the results obtained by the long-term trend analysis of the hydro-climatic variables, the analysis of streamflow response to land use and land cover change, the quasi 3D morphological model and, finally, the ecohydrological analysis, this research has provided in-depth insights and has improved our understanding of the impact of land degradation on the hydrology and morphology of the Dinder and Rahad rivers, and interlinkages to the ecohydrology of the DNP. This is very important for the basin-wide water resources management and sustainable conservation of the Dinder National Park as well as for future research in the D&R basins.

# SAMENVATTING

De stroomgebieden van de Dinder River en Rahad Rivier(D&R) beslaan een grensoverschrijdend oppervlak in Ethiopië en Soedan van ongeveer 77.500 km$^2$. Samen genereert dit oppervlak meer dan 7% van het Blauwe Nijl-water. De Rahad levert water aan het Rahad-irrigatieschema in Sudan (100.000 ha), terwijl de Dinder het ecosysteem van het Dinder National Park (DNP) van ongeveer 10.291 km$^2$ van water voorziet. De overstromingsgebieden van de twee rivieren hebben de afgelopen jaren echter aanzienlijke hydrologische veranderingen ondergaan. Dit heeft grote gevolgen voor het ecosysteem van de zogenaamde "maya's" in het DNP. Maya is een lokale naam voor wetlands van oude rivierbochten die zijn afgesneden van de meanderende rivier die zich aan beide zijden van de Dinder en haar zijrivieren bevinden. Maya's zijn belangrijke ecosystemen in het park, aangezien ze de belangrijkste bron van voedsel en water vormen voor dieren in het wild tijdens het droge seizoen, dat zich uitstrekt van november tot juni. Bij een geschikte ontwikkeling van de watervoorraden en het behoud van een duurzaam ecosysteem moet rekening worden gehouden met de klimaatvariabiliteit en de veranderingen in landgebruik en landbedekking (LULC) en hun effecten op het stroomgebied. In tegenstelling tot de Blauwe Nijl zijn er voor de twee stroomgebieden van Dinder en Rahad maar heel weinig studies uitgevoerd.

De stroomgebieden van de Dinder en Rahad hebben een complexe hydrologie, met een klimaat, topografie, bodem, vegetatie geologie, landgebruik dat sterk varieert. Hoewel het gebied nog een grote variëteit aan natuurwaarden heeft, wordt die bedreigd door schadelijke menselijke activiteiten zoals intensieve begrazing, ontbossing en cultivering van steile hellingen. Deze menselijke praktijken hebben de natuur en plantensoorten in het DNP beïnvloed en vormen een grote bedreiging voor de duurzaamheid en de integriteit van het ecosysteem.

De ruimtelijke en temporele variabiliteit van het hydro-klimaat en veranderingen in landgebruik behoren tot de meest uitdagende problemen. Het begrijpen van de interactie tussen klimaat, Land Use and Land Cover (LULC) veranderingen en hun verbanden met hydrologie, riviermorfologie en ecohydrologie in de Dinder- en Rahadstroomgebieden is moeilijk gezien het gebrek aan klimatologische, hydrologische en ecologische gegevens. Er is slechts een beperkt begrip van de hydrologische processen van het stroomgebied. In het bijzonder over de huidige en toekomstige hydrologische dynamiek is nog weinig bekend. Daarom zijn hydrologische studies van de stroomgebieden van cruciaal belang voor de planning en het beheer van watervoorraden en natuur.

Dit onderzoek analyseert de effecten van landdegradatie op de hydrologie en morfologie van de Dinder en Rahad, en de koppeling met het ecohydrologische systeem van het

Dinder National Park in Sudan. Het gebruikt een ensemble van technieken om ons begrip van de hydrologische processen en LULC-veranderingen in de stroomgebieden te verbeteren. Het onderzoek bevat een lange termijn trendanalyse hydroclimatische variabelen, een analyse van landgebruik en veranderingen in landbedekking, veldmetingen, modellering van neerslagafvoer, GIS en teledetectie data-acquisitie en analyse, hydrodynamische en morfologische modellering van de Dinder-rivier en zijn uiterwaarden, met speciale focus op de maya wetlands. Bovendien is dit onderzoek het eerste onderzoek naar de eco-hydrologie van de DNP. De output van de studie is belangrijk voor het ondersteunen van besluitvormingsprocessen voor een beter beheer van de watervoorraden en het behoud van ecosystemen in het gebied en mogelijk daarbuiten.

De studie beoordeelt de langetermijntrends van de hydro-klimatologie van de Dinder- en Rahad-stroomgebieden werden beoordeeld. Door niet-parametrische Mann-Kendall (MK) en Pettitt-tests worden de trends en de veranderingspunten van hydro-klimatologische tijdreeksen van stroming, regenval en temperatuur geanalyseerd. Trends worden beoordeeld op significantieniveau van 5% voor verschillende tijdsperioden en variërende lengtes op basis van beschikbaarheid van gegevens. De indicatoren voor hydrologische veranderingen benadering (IHA) van Richter et al. (1996) is toegepast om de MK-test te verifiëren en om de essentiële kenmerken van de stroming te analyseren die waarschijnlijk van invloed zijn op ecologische functies in de D & R-stroomgebieden. Deze kenmerken zijn onder andere het grootte, de timing van de afvoer en de snelheid waarmee deze verandert. Het begrijpen in hoeverre afvoeren n zijn veranderd ten opzichte van de natuurlijke omstandigheden is cruciaal voor het ontwikkelen van een effectief beheerplan voor het behoud en herstel van ecosystemen.

Zowel de jaarlijkse afvoer als de seizoensafvoer van de Rahad Rivier vertoont een significant stijgende trend voor de periode 1972-2011. Er was geen waarneembare verandering in de gemiddelde jaarlijkse en seizoensgebonden afvoerpatronen van de Dinder Rivier. De analyse van seizoensmaxima suggereerde echter een verschuiving naar afgenomen afvoer tijdens de periode met hoge afvoer (augustus) en verhoogde afvoer tijdens de periode met lage afvoer (november). De Dinder-maxima van augustus zijn gedaald van 517 $m^3/s$ over het eerste deel van het record (1972-1991) tot 396 $m^3/s$ over de laatste jaren (1992-2011). De gemiddelde jaartemperatuur vertoonde significant stijgende trends met een snelheid van 0,24 en 0,30 $^{o}C$/decennium voor de twee onderzochte stations. Neerslag liet geen significante verandering zien.

De op IHA gebaseerde analyse toont voor de afvoer van de Rahad-rivier significante opwaartse veranderingen in sommige van de hydrologische indicatoren. De afvoer van de Dinder-rivier gaat gepaard met aanzienlijke neerwaartse veranderingen. Het betreft met name: a) een afname van de piekafvoer in augustus en een toename lage afvoeren (november); b) een afname in omvang van extreme afvoeren (d.w.z. 1, 7, 30 en 90 dagen maxima); en c) een afname in snelheid van debietstoename en een toename in snelheid van debietsafname. Deze veranderingen in de Dinder-rivierstromen zullen waarschijnlijk de ecosystemen van het DNP negatief beïnvloeden. De trendanalyseresultaten suggereren dat andere factoren dan klimaatvariabiliteit (bijv. veranderingen in landgebruik en vegetatie) verantwoordelijk zijn voor afvoerveranderingen.

Om de LULC-veranderingen en hun implicaties voor de hydrologie van de Dinder- en Rahad-stroomgebieden te begrijpen, werd een analyse van de debietrespons op landgebruik en veranderingen in vegetatie uitgevoerd met behulp van satellietgegevens en hydrologische modellen. Het hydrologische model van WFlow werd gekalibreerd en gedraaid met verschillende landgebruiks- en landbedekkingskaarten uit 1972, 1986, 1998 en 2011 met vaste modelparameters. De stroomgebiedstopografie, vegetatie en het bodemgebruik werden afgeleid van satellietbeelden en dienden om modelparameters te schatten. De resultaten van de detectie van LULC-veranderingen tussen 1972 en 2011 duiden op een significante afname van bos en een toename van akkerland. Het aandeel bos nam af van 42% tot 14% en van 35% tot 14% voor respectievelijk de Dinder en de Rahad stroomgebieden. Het akkerland nam toe van 14% tot 47% en van 18% tot 68% voor respectievelijk de Dinder en Rahad stroomgebieden. De modelresultaten laten zien dat de afvoer van zowel de de Dinder Rivier als de Rahad Rivier wordt beïnvloed door LULC-veranderingen. De LULC-veranderingen leidden tot een significante toename van de afvoer tussen 1986 en 2011, vooral in de Rahad-rivier. Dit kan worden toegeschreven aan de grote afname van bossen van 35% in 1972 tot 14% in 1986, en de grote uitbreiding van akkerland in het stroomgebied van de Rahad tot 68% van de totale oppervlakte in 2011.

In Dinder nam de jaarlijkse afvoer tussen 1972 en 1986 met 20% toe. Tussen 1986 en 1998 nam de jaarafvoer echter met 9% af. Dit zou het gevolg kunnen zijn van een afname van het bos van 42% in 1972 tot 23% in 1986 en een toename van struikland, grasland en akkerland met respectievelijk 83%, 10% en 6%. In de periode 1986-1998 stegen de akker- en bosgebieden met respectievelijk 192% en 16%, terwijl de overige categorieën daalden. In de periode 1998–2011 nam de jaarlijkse afvoer toe met 52%. Deze toename hangt samen met stijgingen van het percentage kaal land, akkerland en struikland met respectievelijk 360%, 4% en 71%, en afnames van bos en grasland met respectievelijk 50% en 76%. De afname in procentuele verandering van het kale gebied in de periode 1986-1998, samen met de toename van het bos in zowel de Dinder- als de Rahad-stroomgebieden, geeft aan dat de omgeving herstelde van de ernstige droogte van 1984–1985.

Een quasi 3D-model werd gebruikt om de morfologische veranderingen te begrijpen en daarmee de besluitvorming voor het beheer van de maya-ecosystemen te ondersteunen. De analyse richtte zich met name op het effect van morfologische veranderingen op zowel de Dinder-rivier als de maya wetlands. De modelomvang besloeg een oppervlakte van ongeveer 105 km2 binnen het Dinder National Park. SRTM, een digitaal hoogtemodel (90 m DEM) werd gebruikt voor een sectie van 20 km van de Dinder Rivier in Soedan. Aangezien de verticale nauwkeurigheid van de 90 m DEM slecht presteert in gebieden met matige topografische variatie en beboste gebieden, werden in de jaren 2013 en 2016 twee veldtopografische onderzoeken uitgevoerd met behulp van landmeetinstrumenten en geografische informatiesysteem (GPS) apparaten om een DEM met hogere nauwkeurigheid te genereren (verticale fout van 0,008 m en horizontale fout van ± 3 m) voor het modeldomein binnen de DNP. Het gebruik van deze gegevens met een hoge verticale nauwkeurigheidsonderzoek van de uiterwaarden topografie verkregen via de veldonderzoeken maakte de simulatie van het maya wetlands vul- en ledigingsmechanisme mogelijk.

Wegens het ontbreken van waterstandsmetingen binnen het pilotgebied binnen de DNP, is in juni 2013 een meetnet opgezet met zogenaamde Divers om de waterstandgegevens voor dit onderzoek te meten. Het netwerk bestaat uit twee Mini-Divers voor het registreren van waterstand-, temperatuur- en atmosferische drukmetingen en één Baro-Diver om de atmosferische druk te meten die wordt gebruikt om de variaties in atmosferische druk te compenseren. De geregistreerde waterstanden zijn gebruikt om het hydrodynamische model te kalibreren.

Om de hydrologische en morfologische connectiviteit van de maya te begrijpen in termen van vul- / ledigings- en sedimenttransportprocessen, werden zes scenario's geanalyseerd. De eerste drie scenario's simuleren drie verschillende hydrologische omstandigheden van natte, gemiddelde en droge jaren voor het bestaande systeem met het aangelegde toevoerkanaal. De andere drie scenario's gaan uit van dezelfde hydrologische omstandigheden maar voor het natuurlijke systeem zonder toevoerkanaal. De vergelijking tussen scenario's toonde aan dat de hydrodynamica en sedimentologie van de maya worden bepaald door twee factoren: a) de hydrologische variabiliteit van de Dinder Rivier; en b) afgezet sediment bij het inlaatkanaal van het natuurlijke drainagenetwerk.

Ten slotte werd de ecohydrologie van maya wetlands in het DNP beoordeeld en werden relaties tussen vegetatiedynamica, de hoeveelheid wilde dieren en de beschikbaarheid van water geïdentificeerd. Om de ecosysteemstatus en veranderingspatronen te beoordelen, werden veldgegevens over vegetatiesamenstelling en wilde dieren verzameld van vier maya's namelijk; Ras Amir maya, Musa maya, Gererrisa maya en Abdelghani maya. Om de status van het functioneren van de maya's te bepalen, werd een systematic-random quadrat (SRQ) -methode gebruikt voor het verzamelen van gegevens van de flora van de vier maya's binnen het DNP. De genormaliseerde verschilwaterindex (NDWI) werd gebruikt om de inundatiegraad te schatten en de genormaliseerde verschilvegetatie-index (NDVI) werd gebruikt om de gerelateerde vegetatiedekking in de pilot Maya Musa te schatten. Gegevens over de tellingen van wilde dieren in de vier maya's werden geanalyseerd en er werden relaties met hydrologische variabiliteit en vegetatiedekking geïdentificeerd. De SRQ-enquête onderscheidde zeven plantensoorten in de vier onderzochte maya's, met een floristische samenstelling van plantensoorten die aanzienlijk varieert tussen de bestudeerde maya's. De NDVI-analyse van de gegevens tussen 2001 en 2016 toonde significante variaties op het gebied van vegetatiebedekking. Deze variaties waren sterk verbonden met variaties in de NDWI. Uit de tellingen van in het wild levende dieren bleek dat de populatiegrootte en de verspreiding van in het wild levende dieren in het DNP voornamelijk afhangen van de beschikbaarheid van water en graslanden die worden beïnvloed door hydrologische variabiliteit. 84% van de totale populatie wilde dieren (herbivoren) werd gevonden in het grasland in de periferie van maya's, vergeleken met slechts 16% in de verbrande en open gebieden. Dit geeft aan dat herbivoren de voorkeur geven aan grasland en bos rond de maya's in plaats van verbrande en open gebieden. Dit komt waarschijnlijk door de beschikbaarheid van water, voedsel (grasland) en onderdak. Daarom lijkt hydrologische variabiliteit een sleutelfactor te zijn in de ecologische processen.

Gezien de resultaten verkregen door de langetermijnanalyse van hydro-klimatologische variabelen, de analyse van de afvoerrespons op veranderingen van landgebruik en vegetatie, het quasi 3D-morfologische model en, ten slotte, de ecohydrologische analyse, heeft dit onderzoek het begrip van de impact van landdegradatie op de hydrologie en morfologie van de Dinder Rivier en de Rahad Rivier en de verbanden hiervan met de ecohydrologie van de Dinder National Park verbeterd. Dit is belangrijk voor het waterbeheer in het stroomgebied en de duurzame instandhouding van het Dinder National Park alsmede voor toekomstig onderzoek in de Dinder en Rahad stoomgebieden.

het DNP voornamelijk [illegible] van de beschikbaarheid van water en [illegible] worden beïnvloed door hydrologische [illegible] 8 [illegible] van de [illegible] dieren [illegible] [illegible] gevonden in [illegible] [illegible] [illegible] van de [illegible] en [illegible] [illegible] [illegible] [illegible] en bos [illegible] [illegible] [illegible] [illegible] [illegible]. Dit [illegible] ontwikkelt. [illegible] [illegible] onderzoek. Daarom zijn hydrologische [illegible] ecologische processen.

Om [illegible] de [illegible] [illegible] [illegible] [illegible], de [illegible] [illegible] [illegible] [illegible]

[illegible]

# CONTENTS

# 1

# INTRODUCTION

## 1.1 Background

In the face of declining water resources on the global scale, the international scientific community has emphasized the need for new solutions to address the global water crisis. There is growing awareness that integrated water resources management is required, because freshwater resources are limited and becoming more and more unfit for human consumption and also unfit to sustain the ecosystem (Savenije and Van der Zaag, 2008).

Hydrology is recognized as a critical factor in the preservation of the ecosystem integrity of streams and rivers. The understanding of the relationships between the flow regime of a river and its ecological functioning is crucial for developing appropriate techniques to manage ecosystem integrity. In addition, as part of integrated water resources management focus also needs to be on maintaining and restoring ecosystems health and biodiversity (Jewitt, 2002).

The research on the interaction between hydrological and ecological systems relates to different levels and scales. A number of studies present an increasing linkages between hydrology and ecology in various fields of research, such as ecohydrology (Richter et al., 1996; Wassen and Grootjans, 1996; Gurnell et al., 2000; Zalewski, 2002; Kundzewicz, 2003; Baird et al., 2004; Hannah et al., 2004) or riverine landscape ecology (Poole, 2002; Stanford, 2002; Tockner et al., 2002; Ward et al., 2002; Wiens, 2002; Schröder, 2006). With time, ecohydrology emerged as a new interdisciplinary field or even a paradigm (Bond, 2003; Hannah et al., 2004; Rodríguez-Iturbe and Porporato, 2005).

Ecohydrology is an important concept that is built on the capability of science to describe and quantify the relationships between hydrological processes and biotic dynamics at basin scale and, if required, to employ these processes to increase the strength of the aquatic system and thus its capacity to cope with human induced pressures (Hu et al., 2008). This approach necessitates a sound knowledge of ecosystem functioning, as a basis for enhancement of the interaction between hydrologic and ecological factors.

McCalin et al. (2012) underlined that ecohydrology is a trans-disciplinary science originated from the larger earth system science frameworks and examining common connections of the hydrological cycle and biological communities and is becoming a quickly developing branch of knowledge in hydrological science. It is likewise a connected science concentrated on critical thinking focused on problem solving and giving sound direction to basin-wide integrated land and water resources management.

Zalewski (2002) defines ecohydrology as 'the study of the functional interrelations between hydrology and biota at the catchment scale' and 'a new approach to achieving sustainable management of water' and views the research field mainly as dealing with aquatic systems. This definition has widespread applicability, as it recognizes the two-way interaction between hydrology and ecology (Wood et al., 2008). Nevertheless, this definition has been adopted mainly within the framework of water resources management and biological conservation in terms of assessment of ecosystem reactions to natural and human induced water stresses (Zalewski, 2002).

Many rivers in the world have suffered a long history of degradation through direct or indirect human interventions (Maddock, 1999). The magnitude and velocity of water movements through river channels, its floodplain and surface water and groundwater interactions have been changed through the impacts of climate change and land use land cover change. The negative effects of these impacts from a conservation viewpoint have been extensively reported (Maddock, 1999). If a wetland area is lost, related ecosystem processes and services are also lost. However, it is important to note that wetlands provide significant global ecosystem services such as biodiversity support, food for a range of living beings, water quality improvement, flood retention and carbon management. Each of these services depend on a range of bio-physical interactions.

Over the past decades, identification of the adverse consequences of both human and natural impacts on rivers, combined with an increase in overall environmental awareness, guided to many initiatives for river restoration as part of river basin management programs. Some river restoration studies intended to enhance the water quality (Jordan et al., 1990) while others intended to enhance the ecological integrity of river systems (RRP, 1993). No matter what the driving force are, there is a developing scientific knowledge related to theories, methods and effective applications of river ecosystem restoration being applied over the world (e.g. Brookes and Shields, 1996; Connelly and Knuth, 2002; Giller, 2005; Wohl et al., 2005; Kondolf, 2006; Palmer et al., 2010; Bernhardt and Palmer, 2011). The role of streamflow and the river channel morphology in defining the structure of river ecosystems received little consideration until the early 1980s (Newbury, 1984; Nowell and Jumars, 1984). Maddock (1999) emphasized that upcoming studies on the growth of physical habitat assessments must attempt to integrate and combine the wide range of spatiotemporal scales that affect the ecosystem functioning and hence the human wellbeing.

Many wetlands around the world endangered by alterations in hydrological regime or land use and land cover changes, require efficient management policies and practices to conserve them (Alvarez-Mieles et al., 2013). In spite of the importance of the Dinder and Rahad (D&R) basins for bio-physical as well as human systems in the region, only few scattered studies on climate and wildlife conservation and management have been carried out in these basins. The hydrology, the land use and land cover (LULC) changes and the ecohydrology of the basin have not been studied and understood. Accordingly, it is very important to study the relationship between hydrological and ecological processes and patterns and the interaction between LULC changes, hydrology, river morphology and the ecohydrology of the mayas wetlands inside the DNP. Maya is a local name for floodplain wetlands and oxbows cut off from the meandering river that are found on both sides of the Dinder river and its tributaries (Hassaballah et al., 2019).

## 1.2 MAYAS ECOSYSTEM MANAGEMENT

The aim of the mayas ecosystem management is to sustain the DNP ecosystem integrity by protecting the indigenous biodiversity and the ecological evolutionary processes that create and maintain that diversity.

There are a large number of examples presenting the influence of hydrologic regime on ecological process and patterns and riverine landscapes (Schröder, 2006). As an example, Naiman and Decamps (1997) along with Ward et al. (2002) assessed the ecological diversity of riverine landscapes. In such case, the changing environment support organism's adaptation to disrupted regimes over wide spatiotemporal scales (Lytle and Poff, 2004). Robinson et al. (2002) reported that the movement of many species is strongly linked to the spatiotemporal dynamics of the shifting landscape ecology. Tabacchi et al. (1998) assessed how vegetation dynamics are affected by the hydrological alterations and, on the other hand, how vegetation diversity and productivity influence riverine geomorphologic developments. Recently, Berhanu and Teshome (2018) reported that Alatish National Park inside Ethiopia, on the other side of the border with DNP, was badly degraded due to shortage of seasonal water among other factors.

Another example presenting the effects of hydrologic patterns and processes on ecological features refers to stream channel modification that occurred either naturally through erosion and sedimentation processes, or man-made channelization. Channel modification may cause significant change of the magnitude and duration of flooding and sedimentation. The alterations in the hydrogeomorphological process prevent/support the river and the floodplain interaction, which in many cases alter the composition of plant communities (Shankman, 1996). The clear example for channel modification by human that is relevant for the study area, is the canalization of mayas' feeders inside the DNP by creating canals to divert water from the river into the mayas during the flood season.

Similarly, literature showed many examples describing the influence of ecological processes and patterns on hydrological processes. Tabacchi et al. (2000) analyzed the effects of riparian vegetation on hydrological processes, i.e.: (a) the effect of plant growth on water uptake, storage capacity and return to the atmosphere, (b) the control of runoff by the physical influence of living and dead plants on hydraulics, and (c) the effect of riparian vegetation functioning on water quality. Mander et al. (2017), demonstrated how the potential hydrological returns from investing in ecological infrastructure can be modelled. Their research concluded that considerable benefits in both water quantity and quality could be achieved with interventions to rehabilitate and maintain water-related ecological infrastructure at a catchment scale to improve water security.

Another example describing the effects of ecological processes and patterns on hydrological regime refers to the so-called ecosystem engineers (Jones et al., 1994; Jones et al., 1997; Alper, 1998; Bruno, 2001; Crain and Bertness, 2006; Hastings et al., 2007; Wright, 2009; Jones et al., 2010). The terms "Ecosystem engineering" which refer to the process, and "Ecosystem engineers" which refer to the organisms responsible, were originally proposed by Jones et al. (1994). Ecosystem engineers defined as an organism that modify, maintain and/or create habitat. Ecosystem engineering leads to changes in two ways. First, through "autogenic engineering" in which the structure of the engineers itself alters the environment (e.g. tree growth) and the engineer remains as part of the engineered environment. Second, through "allogenic engineering" in which organisms transform habitats or resources from one physical state to another and the engineer is not necessarily part of the permanent physical ecosystem (e.g. beaver dams). Both animals and plants can be both autogenic and allogenic engineers (Jones et al., 1997). Such processes retain sediments and organic matter in the channel, influence the structure and dynamics of the riparian zone, change the characteristics of water and materials transported downstream, modify nutrient cycling and eventually influence plant and animal community composition and diversity" (Naiman et al., 1988). Understanding the ecosystem engineering processes required empirical data from comparative and experimental studies, models and conceptual integration of the processes (Jones et al., 1997), which are not available for the DNP. Thus, studying the ecosystem engineering process is beyond the scope of this research and is not part of our analysis.

## 1.3 PROBLEM DESCRIPTION

The hydrological processes of the Dinder and Rahad basins are not well understood, in particular the prediction of hydrological and morphological dynamics have not been studied before. Therefore, in-depth hydrological and morphological studies of the basins and their interactions with the ecosystem are very essential to inform better understanding and management of water resources and ecosystem of the D&R basins. The ecologically wealthy DNP depends mainly on the ecosystem services provided by the mayas particularly in the dry season which extent from November to June. During the past three decades, the area of some mayas inside DNP have radically decreased. Such mayas can no longer store sufficient water to meet the requirements of the wildlife populations in the park throughout the dry season. Some mayas were completely dry up (Figure 1.1), and the causes are not understood. The drying of mayas could have serious impacts on wildlife populations that depend on them for water and food in the dry seasons. Thus, the entire ecosystem of the DNP seems vulnerable to hydrological and morphological changes because it largely depends on the mayas. Although many of the mayas have degraded (e.g. Musa maya, Ein Elshamis, Biet Elwahash and Gererrisa), there is no evidence of new mayas being formed, since the establishment of the DNP in 1935. To sustain the ecosystem, some mayas are artificially watered from boreholes drilled near the mayas, and some of them are watered during the wet season from the Dinder river by artificial canals. In many situations, this engineering approach has led to significantly engineering of the environment and ecosystem processes and services. This seriously impacts the role of the ecological processes in moderating the water cycle and sediment dynamics.

Therefore, assessments of water resources through data collection and hydrological and morphological models, analysis of land use land cover changes as well as ecohydrological analysis are important for filling important knowledge gaps related to the conservation of the DNP and gaining new insights into the hydrology of the Dinder and Rahad basins and better understanding of the key factors affecting the functioning of mayas inside the DNP.

*Figure 1.1: Dry mayas in the Dinder National Park. (Pictures taken by Khalid Hassaballah, March 2011).*

## 1.4 RESEARCH OBJECTIVES

This research aims to improve the understanding of the interactions between hydrology, river morphology, land degradation and ecohydrology of the Dinder and Rahad river basins to support conservation and sustainable management of the ecosystem. Although, sustainable catchment management and ecosystems conservation requires integration of the hydrologic, environmental and socioeconomic components that occur within the catchment, socioeconomic processes were beyond the scope of this research. The specific objectives of this research are:

- ✓ To quantify the long-term trends of the hydroclimatic variables in the Dinder and Rahad river basins and assess whether possible trends and changes have affected the functioning of mayas;
- ✓ To evaluate the effects of the land use and land cover changes on the Dinder and Rahad streamflow response;
- ✓ To understand the functioning (i.e., filling and emptying) of mayas and related hydrological and morphological processes along the river reach and within the mayas; and
- ✓ To assess the hydrological controls on vegetation dynamics and wildlife in the maya wetlands of the DNP.

Conducting research on the ecohydrology and morphology of the maya wetlands system and assessing the interrelations with the relevant ecosystem contributes to fill an important knowledge gap on the Nile ecohydrology in this "forgotten" region.

## 1.5 RESEARCH HYPOTHESES

The following hypotheses were proposed to guide this research:

i. The long-term trends in climatic variables are the major drivers for hydrological changes at the river basin scale.

ii. Land use and land cover changes are the main drivers of changes affecting the hydrology of Dinder and Rahad basins.

iii. Hydrological alterations and morphological changes are the main factors controlling the filling mechanism of the maya wetlands of the DNP.

iv. Water availability is the main factor affecting the vegetation dynamics and wildlife population in mayas ecosystem of the DNP.

## 1.6 RESEARCH QUESTIONS

A number of research questions were addressed related to the understanding of the interactions between the Dinder and Rahad hydrology, land use and land cover changes, morphological changes and the ecosystem of the DNP. These questions include, most importantly:

1. Is there any significant long-term trend in the hydroclimatic variables in both river basins, and if so to what extent?
2. What are the impacts of the land use and land cover changes in the upper Dinder and Rahad on the catchment runoff response?
3. How does filling and emptying of mayas normally occur, and what are the key factors controlling these processes?
4. Can identified changes of the mayas functioning (i.e., filling and emptying) be related to local ecosystems (e.g. flora and fauna)?

## 1.7 METHODOLOGICAL FRAMEWORK

This research attempted to use a comprehensive approach to examine the hydrological and morphological changes as well as the LULC changes and assess the implication on the ecosystems of the DNP using a range of methods, including statistical analysis of historical data, field observations, GIS and remote sensing data analysis as well as hydrological and morphological modelling. First, statistical tests were used to assess the significance of trends of key hydro-climatic parameters over different time periods. Second, Wflow hydrological model was used to analyze streamflow response to land use and land cover changes. Third, a quasi 3D modelling was used to quantify the effect of morphological changes on both the Dinder river and the maya wetlands. Finally, relations between vegetation dynamics, wildlife and hydrological variability were assessed in maya wetlands along the Dinder river.

The main methodology and activities that were implemented to achieve the research objectives are presented in Figure 1.2. Further details of methods are given in the respective chapters.

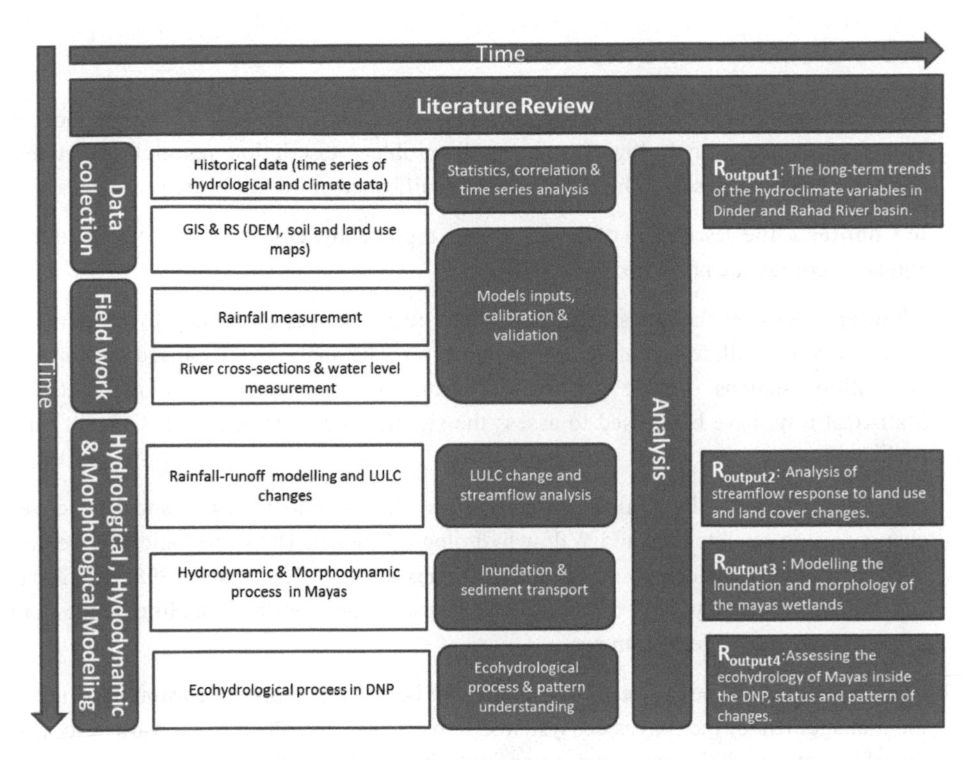


*Figure 1.2: Framework of methods and activities in this PhD research.*

## 1.8 DISSERTATION STRUCTURE

The research is organized in seven chapters. In the first chapter, the importance of hydrological and ecohydrological science, the problem description, research objectives, research questions, research hypotheses and significance of the study are presented.

In **Chapter 2**, the description of the study area, topography, climate, hydrology, land use, soil and geology are presented.

**Chapter 3** presents the assessment of the long-term changes of the key hydro-climatic parameters (rainfall, temperature and streamflow). The analyses are carried out for two streamflow stations, twelve precipitations and two temperature gauging stations. Statistical tests have been used to assess the significance of trends over different time periods.

**Chapter 4** presents the analysis of streamflow response to land use and land cover changes using satellite data and Wflow hydrological model. The hydrological model has been derived by different sets of LULC maps from 1972, 1986, 1998 and 2011. Catchment topography, land cover and soil maps, are derived from satellite images and serve to estimate model parameters.

**Chapter 5** presents how a quasi 3D model can be used to support decision making for the management of the mayas ecosystems. In particular, the annual flow variability and the effect of morphological changes on both the Dinder river and the maya wetlands. Delft3D was applied to a 20 km reach of the Dinder river between Gelagu camp and up to few kilometers downstream of the pilot Musa maya. The discharge data which were computed using the hydrological model presented in (Chapter 4) were used as an upstream boundary condition for the model domain.

In **Chapter 6** the relations between vegetation dynamics, wildlife and hydrological variability were assessed in four mayas along the Dinder river. Field data on vegetation composition and wildlife were collected from the four mayas to assess the ecosystem status and patterns of changes. Relations between hydrological variability, vegetation cover and wildlife populations were identified.

Finally, **Chapter 7** summarizes the main findings of the dissertation and presents some conclusions and recommendations.

# 2

# THE STUDY AREA: DINDER AND RAHAD RIVER BASINS

---

[2] This chapter is based on but not limited to: Hassaballah, K., Y. A. Mohamed and S. Uhlenbrook.: The Mayas wetlands of the Dinder and Rahad: tributaries of the Blue Nile Basin (Sudan). The Wetland Book: II: Distribution, Description and Conservation. C. M. Finlayson, G. R. Milton, R. C. Prentice and N. C. Davidson. Dordrecht, Springer Netherlands: 1-13, 2016.

## 2.1 THE DINDER AND RAHAD RIVER BASINS (D&R)

The Dinder and the Rahad are the lower sub-basins of the Blue Nile river basin located between longitude 33°30' E and 37°30' E and latitude 11°00' N and 15°00' N (Figure 2.1). The Blue Nile basin collects flows of eight major tributaries in Ethiopia besides the two main tributaries in Sudan: the Dinder and the Rahad rivers. Both tributaries receive their water mainly from the runoff generated in the Ethiopian highlands approximately 30 km west of Lake Tana (Hurst et al., 1959). The Dinder river joins the Blue Nile at the village Al-Rabwa, 64 km downstream of Sennar reservoir, while the Rahad river joins the Blue Nile at the village Abu Haraz below Wad Medani town. The D&R generate around 7% of the Blue Nile basin's annual flow. The Rahad river supplies water to the Rahad irrigation scheme (126,000 ha), while the Dinder river supplies water to the diverse ecosystem of the Dinder National Park (DNP). The catchments areas about 34,964 and 42,540 km$^2$ for the Dinder and the Rahad, respectively, resulting in a total area of about 77,504 km$^2$. However, in the Ethiopian highlands where rainfall is relatively high (about 1,400 mm/y), the catchment area of Dinder (18,000 km$^2$) is two times that of the Rahad river (8,758 km$^2$). The total catchment has varied topography with elevation ranging between 384 m at the catchment outlet and up to 2,731 m at the Ethiopian plateau (Figure 2.1). The basin boundary and the streams network have been delineated from a 90 m x 90 m digital elevation model database of the NASA Shuttle Radar Topographic Mission (SRTM) acquired from the Consortium for Spatial Information of the Consultative Group for International Agricultural Research (CGIAR_CSI) website (http://srtm.csi.cgiar.org).

The main soil types in the D&R according to the Food and Agriculture Organization (FAO) classification are: vertisols 71%, luvisols 9%, nitisols 8%, leptosols 5%, cambisols 4%, alisols 2% and fluvisols 1%. The vegetation cover is characterized by grasslands, shrublands, croplands and woodlands.

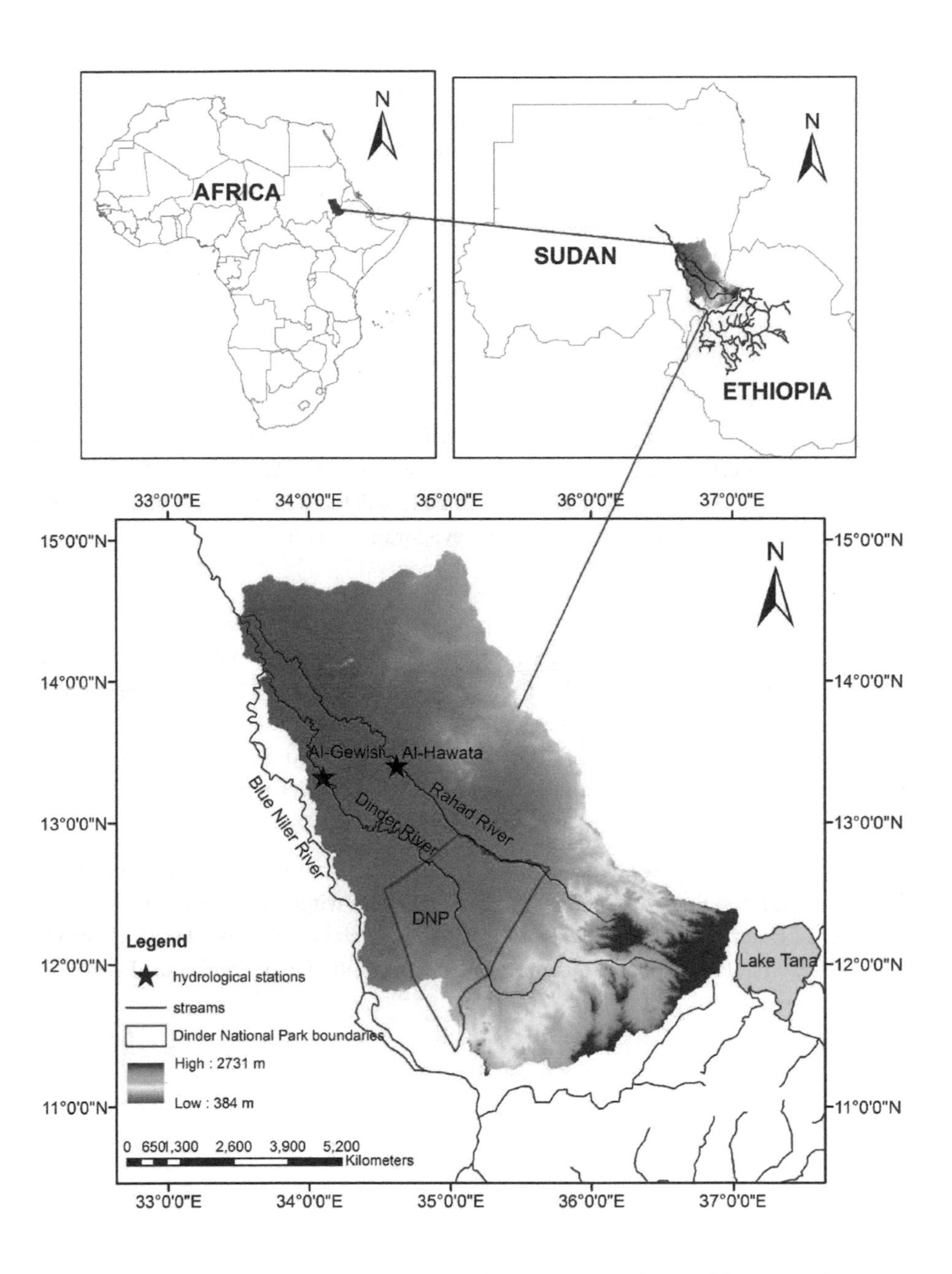


*Figure 2.1: Location and topography map of the Dinder and Rahad basins and the DNP. The two black stars are the hydrological stations (Al-Gewisi and Al-Hawata).*

### 2.1.1 Hydrology and climate

The hydrology of the D&R is complex, with varying climate, topography, soil, vegetation and geology as well as human interventions (Hassaballah et al., 2016). The annual average flow (1900-2016) is about 2.70 x $10^9$ $m^3/a$ and 1.102 x $10^9$ $m^3/a$ for the Dinder and the Rahad, respectively, with the maximum flow during August/September. The monthly rainfall records indicate a summer rainy season with highest total rainfall in the months from June to September (Block and Rajagopalan, 2006). The rainfall during this season accounts for nearly 90% of total annual rainfall in the lower part of the basin, while in the Ethiopian highlands, approximately 75% of the annual precipitation falls during this rainy season (Shahin, 1985).

Regarding the seasonal response of the Dinder and Rahad rivers during high and low precipitation, both rivers are completely dry during the dry season. In the upper part, the rivers are very steep and the numbers of tributaries are high. As the flow of the Dinder river is seasonal, large areas of mayas used to be inundated each year and then dry up as the water infiltrates (groundwater recharge), evaporates or consumed by wildlife.

### 2.1.2 Rainfall

The rainfall accounts to 1400 mm/a in the Ethiopian highlands near Lake Tana and reduces to 900 mm/a at the highland plateaus at the upper part of Dinder and Rahad catchment. In the middle course as at Gelagu station (inside the DNP), the mean annual rainfall is less than 600 mm/a and further in the lower course (in Sudan) it is less than 400 mm/a at the village El Rabwa at the confluence of the Dinder river with the Bule Nile river.

Figure 2.2 shows the variations in the monthly mean rainfall at the Dinder station downstream of the Dinder catchment, the Gelagu station within the mayas area inside the DNP and at Bahir Dar station further upstream of the catchment (Lake Tana). Bahir Dar is the nearest rainfall station to the upper catchment of the Dinder and Rahad with long historic records.

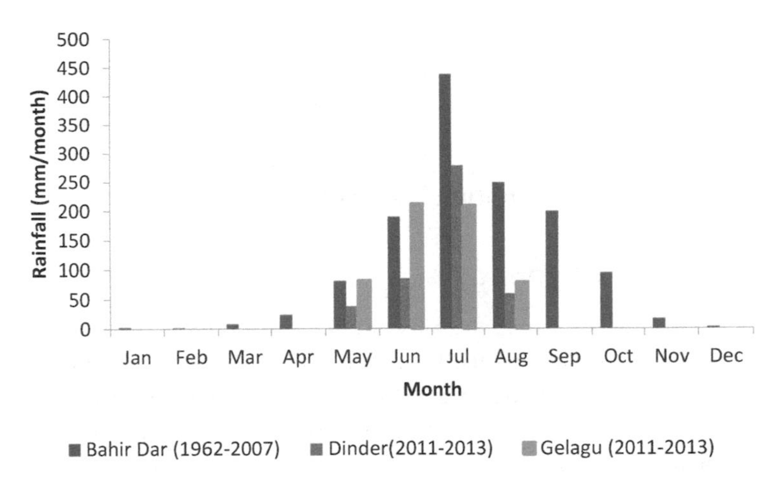


*Figure 2.2:The monthly mean rainfall for Bahir Dar (1962-2007), Dinder (2011-2013) and Gelagu (2011-2013).*

## 2.1.3 Evapotranspiration

The mean annual potential evapotranspiration (PET) follows a similar trend as that of temperature. In the highlands plateau, the PET rate is estimated to be 1320 mm/a. The low-lying area (below 1500 masl) located at the foot of the highland plateaus, up until the border and a little beyond (30% of the Dinder and Rahad basins), experiences mean annual PET that ranges from 1800 to 2280 mm/a (Woolf et al., 2015). Further in the Sudan lowland area at Gelagu station, PET is some 2300 mm/a and further downstream at Dinder is estimated to exceed 2500 mm/a (Block et al., 2007).

## 2.1.4 Temperature

Temperature at the highland plateau of the sub-basin is pleasant and the mean annual temperature does not exceed 20 °C. Large proportions of this highland exhibit mean annual temperatures of 18 °C. In the western low-lying area of the sub-basin, around the border, mean annual temperature is in the order of 25 °C. Further in the downstream part of the sub-basin, around the Gelagu station, mean annual temperature is estimated to be 27 °C. In the lower course, at the mouth of the sub-basin, temperature exceeds 30 °C.

### 2.1.5 Humidity

Nearly 80% of the sub-basin is identified to have a mean annual relative humidity of less than 55%. It is only 20% of the sub-basin with relative humidity exceeding 55%. This portion of the sub-basin is confined in the Ethiopian Plateau.

## 2.2 Dinder River and Dinder National Park (DNP)

### 2.2.1 Dinder river hydrology

The Dinder river originates from the west of Lake Tana in Ethiopia flowing westwards across the Sudan border joining the Blue Nile below Sennar at the village El Rabwa approximately 115 km downstream of Al-Gewisi town. The Dinder river basin has a complex hydrology, with varying climate, topography, soil, vegetation cover and geology as well as human activities.

The Dinder river has a length of about 750 km with no large tributaries except Khor Gelagu and Khor Masaweek inside the DNP, Khor Kenana and Khor Abu Muhar on the left bank of the river and Khor Abu Al-Hasan on the right bank, which connects to the Dinder river a few kilometers upstream of Al-Gewisi station (Figure 2.3). The Khor is a seasonal or dry watercourse. During extream flood events in the Rahad, Khor Abu Al-Hasan diverts water from the Rahad river to the Dinder river. All Khors are completely ungauged. The Dinder river loses some of its water in swamps along its course and by spilling on both the left and right banks of the river.

The flow records of the Dinder between (1972-2015) show an annual average flow of about 2.20 $x10^9$ $m^3/a$. Figure 2.4 shows the high variations in the annual and daily flows of Dinder during the period 1972-2015. Considering the seasonal flow behavior of the Dinder, the river carries a considerable discharge in only four months of the year (July-October). The flow duration extends from July to December. For about a half year, from January to June the sandy bed of the river is left with only few pools which may hold water until the next rainy season.

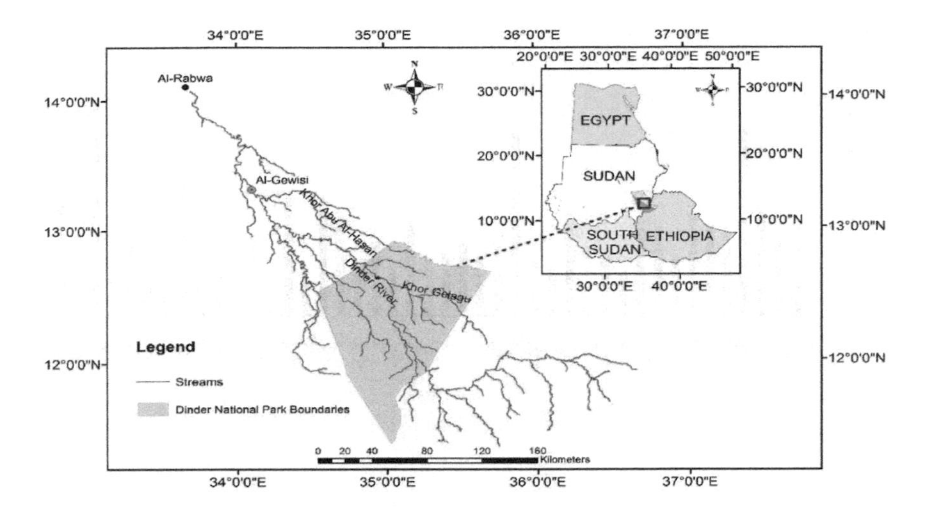


*Figure 2.3: The study area of the Dinder river basin and the Dinder National Park (DNP).*

(a)

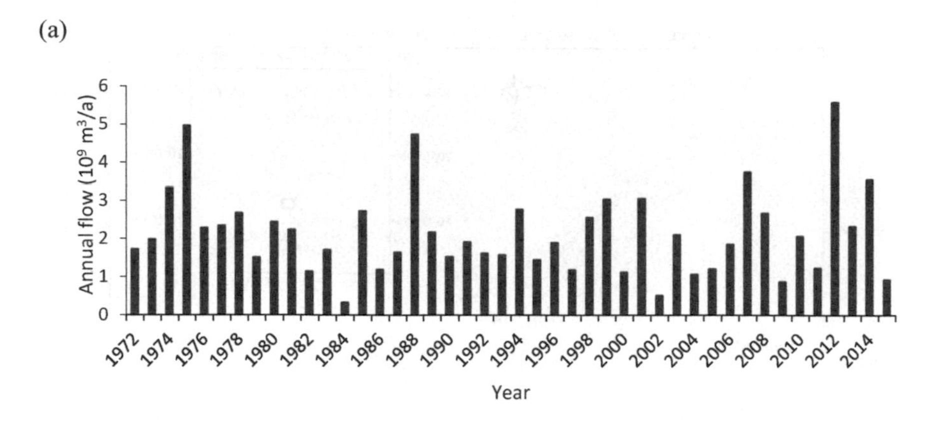


(b)

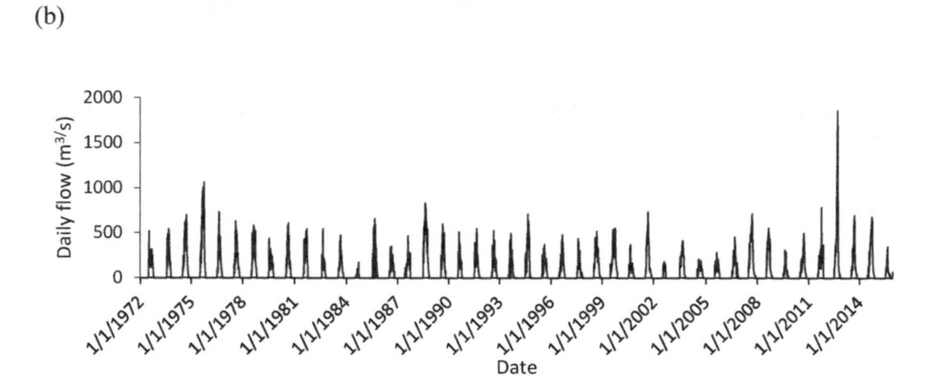


*Figure 2.4: The Dinder river flow at Al-Gewisi station (a) the annual flow and (b) the daily flow.*

The Dinder river records (Figure 2.5) indicate a reduction of flows in recent years. The annual average flow recorded at the mouth of the Dinder (1900-1960) at Hillet Edreis station is 3.05 x $10^9$ m$^3$/a, compared with the later record (1961-2016) of 2.33 x $10^9$ m$^3$/a at El-Gewisi station. This comparison underestimates the actual decline of flow because channel losses between the upper and lower sites are ignored. The long-term annual average flow (1900-2016) is about 2.70 x $10^9$ m$^3$/a which represents over 5% of the Blue Nile basin's annual flow. The river starts flowing in July and reaches its peak flow in August-September, with a short low flow recession period from October to November. The range of annual flows is large. The maximum recorded in the early years was 5.64 x $10^9$ m$^3$/a in 1916, compared with low flow of 1.24 x $10^9$ m$^3$/a in 1941. This low flow record has been superseded in 1984 by flow of only 0.31 x $10^9$ $m^3$/a. There is an extended dry period from January to May/June when flow is zero in most years.

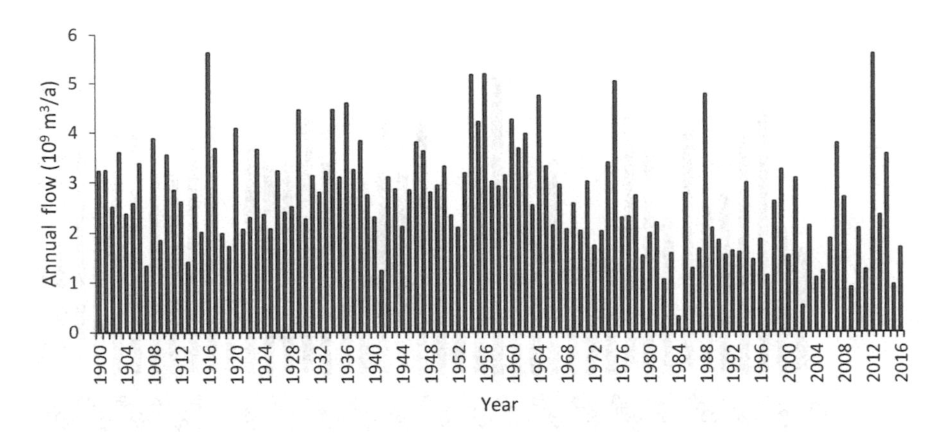


*Figure 2.5: Annual flow of the Dinder river from 1900-2016 (Source: ENTRO and MoIWR).*

The river is the main source of water for the diverse ecosystem of the DNP. Although the Dinder flow is seasonal, the areas of maya wetlands used to be inundated during the high flood periods depending on the magnitude of flood. Complicating the mayas hydrology is also the man-made canalization of feeders (small canals that supply water to mayas) to enhance the filling of mayas (Hassaballah et al., 2016).

The river also supplies water to the small-scale horticulture activities (Jeroof) of local communities on both sides of the river and inundate large areas of floodplains outside the DNP which provide drinking water storage for both domestic use and nomadic livestock during the dry period. During the dry season the river bed is left with small pools which were consumed by the wildlife in the river reach inside the DNP.

## 2.2.2 Rahad river hydrology

The Rahad is a tributary of the Blue Nile on the right side originates from Ethiopia. The Rahad river has a length of about 800 km with few tributaries (Khors) such as: Khor Abu Farga, Khor Samsam, and Khor Almasub with flash floods. The total catchment area is about 42,300 $km^2$ and its effective catchment in Ethiopia is about 8,200 $km^2$ and its annual average flow is about 1.10 x$10^9$ $m^3$/a with a maximum ten day mean of 160 $m^3$/s. All Khors are ungauged, however, the water balance computation has shown that the annual average flow of Khor Abu Farga (2015-2019) is about 0.55 x$10^9$ $m^3$/a. Figure 2.6 shows the annual flow of the Rahad river at Al-Hawata station.

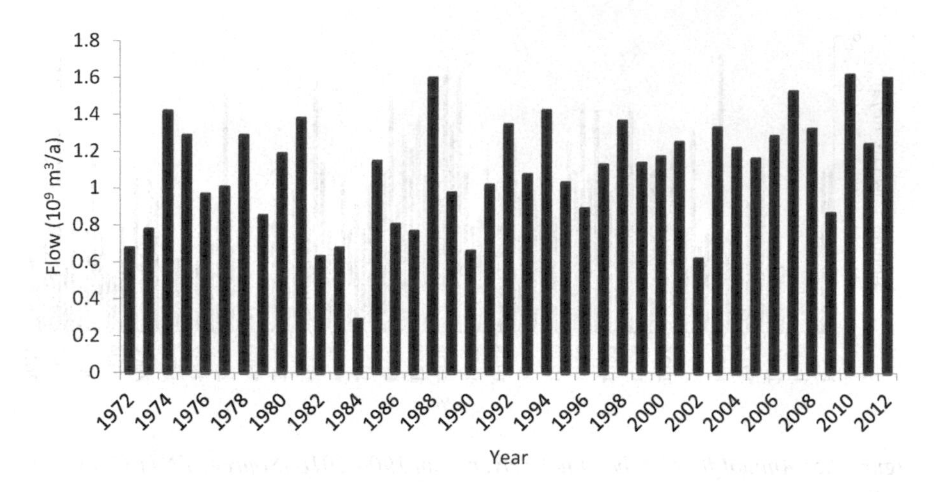


*Figure 2.6: Annual flow of the Rahad river at Al-Hawata station.*

The Rahad loses much of its total water in swamps along its course and by spilling on the left bank and some of its water reaches the Dinder river through Khor Abu Al-Hasan and some infiltrate to recharge the groundwater. Its total contribution to the Blue Nile is only one third of the Dinder.

The average annual flow of the Blue Nile (1912-2015) at El Diem station near to the Sudan border with Ethiopia is about 49 x $10^9$ $m^3$/a; the daily flow fluctuates between 10 x $10^6$ $m^3$/day in April to 800 x $10^6$ $m^3$/day in August.

The only major water infrastructure on the Rahad River is Abu Rakham barrage (Figure 2.7), which serves as the major regulator of both the supply from Rahad River and the Mena pumping station from the Blue Nile River. This is a diversion structure constructed on the Rahad River some 47 km downstream of Al-Hawata gauging station and about 145 km upstream of its junction with the Blue Nile River. The main purpose of the barrage is to divert water from the Rahad River to the Rahad scheme during the flood season. Fifteen vertical sluice gates are constructed for the operation of the barrage across the Rahad River. Nine gates operated on the Rahad River, while six gates are operated on the main irrigation canal. Dimensions of the gates are 6 m in length and 4 m in width. The height of the barrage is approximately 4.5 m.

### 2.2.3 Human interventions in Dinder and Rahad basins

Many areas have been degraded as a result of rain-fed farming and removal of tree cover in the upstream parts. Information on water abstractions in the upper part of the basins in Ethiopia are not available. However, a detailed analysis of the LULC changes in the basins was provided in chapter 4 and no indications for larger abstractions (irrigation schemes, industries etc.) were found.

The only large-scale irrigation project in the basins is the Rahad scheme located on the eastern bank of the Rahad River in the most downstream part of the basin (Figure 2.7). The 126,000 ha Rahad scheme (300,000 Feddans) was planned during the mid-1960s. Execution period began in 1973 and lasted up to 1977 when part of the Rahad scheme was put under cultivation. The whole scheme was fully operated in 1981 (Ibrahim et al., 2009). The water supply sources for the Rahad scheme are the Blue Nile River and the Rahad River. The Mena pumping station located some 75 km upstream of Sennar dam, diverts water from the Blue Nile River to the Rahad scheme through the Rahad supply canal which passes underneath the Dinder river through the Dinder syphon. The capacity and number of pumps (eleven electrical centrifugal pumps) were designed according to the estimated maximum crop water requirements (CWR) during the cultivation season. The peak CWR in the Rahad scheme is 28 $m^3$/Feddan/day, and this figure is used in the design of irrigation networks. The design capacity of main canal is 8.4 million $m^3$/day. The capacity of the pumps is 9.55 $m^3$/sec for each. The total discharge generated by ten operating pumps for 24 hrs (design capacity) is 8.25 million $m^3$/day; one pump is left as a reserve. During the wet season (July-October), the supply to the Rahad scheme is augmented from the Rahad River. The annual average water supply to the scheme during the periods (2000-2004) and (2015-2019) is about 950 x $10^6$ $m^3$/a and 822 x $10^6$ $m^3$/a, respectively. 45% of the water supplied to the scheme is diverted from the Rahad River at Abu Rakham Barrage during the wet season and 55% is diverted from the Blue Nile River from Mena pumping station when the Rahad is dry.

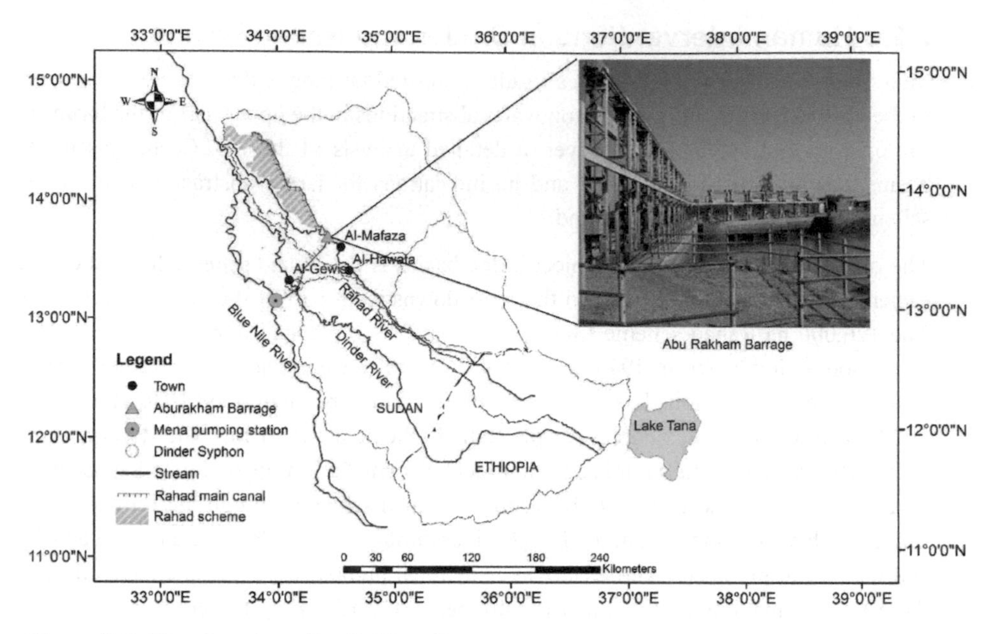


*Figure 2.7: Regulators at Abu Rakham Barrage*

## 2.2.4 The Dinder National Park (DNP)

The DNP (10,291 km$^2$) is a very important ecological area in the arid and semi-arid Sudan-Saharan region. It was proclaimed as a national park in 1935 after the 1933 London Convention for the Conservation of African Flora and Fauna (Dasmann, 1972). It was then declared a biosphere reserve in 1979 (Abdel Hameed, 1998), and registered as a Ramsar site in 2005. The park is located in the South-east of Sudan near the Ethiopian border between latitudes 11°00' and 13°00' N and longitudes 34°30' and 36°00' E (Figure 2.8). The water system of the park depends on both the Dinder river in the middle of the park and the Rahad river on the Northern border of the park and their tributaries and mayas. "Maya" is a local name for floodplain wetlands and oxbow cut off from the meandering river that are found on both sides of the Dinder river and its tributaries.

Mayas are important ecosystems in the park as they constitute the main source of food and water for wildlife during the dry season which extends from November to June. Mayas are generally having crescent shapes. Their areas vary significantly from about 0.16 km$^2$ to 4.5 km$^2$ (Hassaballah et al., 2016). They are normally flat with slight and/or no clear banks. Some of the flooded mayas start to dry up as the water infiltrates (groundwater recharge), evaporates or consumed by wildlife, while others retain water throughout the year. Some mayas have relatively well-defined channels while others do not. Depending on the type and condition of vegetation and the amount of open water, evaporation rates will vary greatly. To sustain the mayas ecosystem, some of the dry mayas are artificially kept wet by pumping groundwater. Complicating the mayas hydrology is also the man-made canalization of feeders (small canals that supply water to the mayas from the river) to enhance the filling of the mayas (Hassaballah et al., 2016).

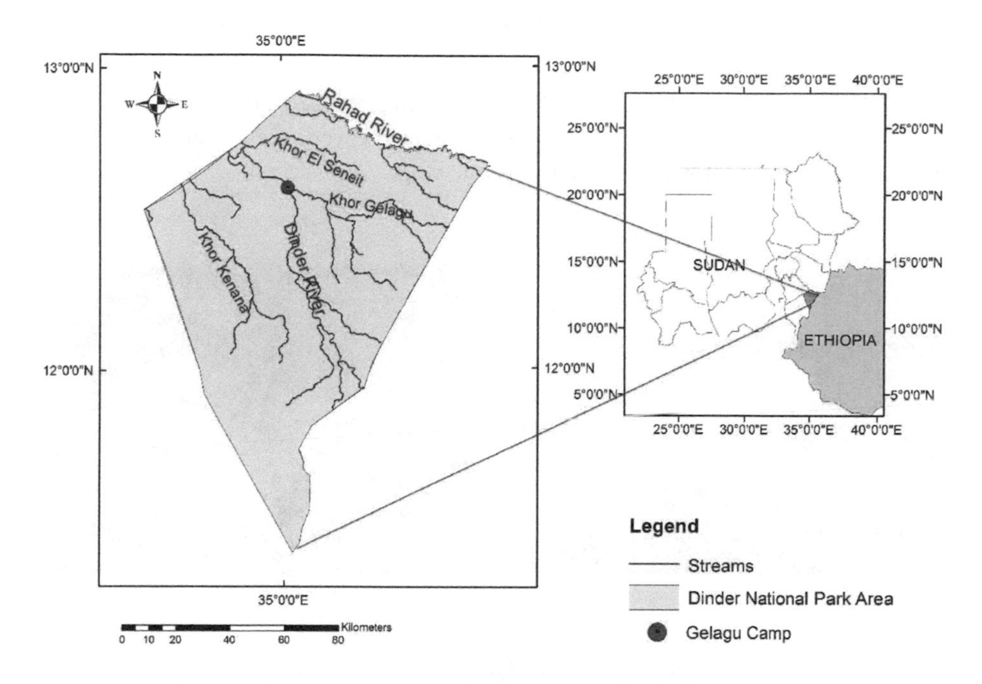


*Figure 2.8: Location and water system map of the Dinder National Park.*

According to DNP authority, there are more than 40 mayas and pools that are part of the Dinder river system. Mayas are the main source of food and water for wildlife (herbivores) in the park specially during the dry season, which extend from November to June. Each maya is a habitat for species that varies in both quantity and quality. In relation to their carrying capacities and water retention potential, mayas are classified as productive and non-productive mayas. The non-productive mayas are the degraded mayas in which the mat-forming grasses were replaced by tall unpalatable seasonal grasses such as *Sorghum sp.* which dry up and are subject to burning (AbdelHameed et al., 1997).

The edges of mayas are surrounded by trees in a consistent pattern (Figure 2.9). Starting from the periphery of the mayas, the vegetation bands consist of *Balanites aegyptiaca, Acacia seyal, Acacia siberiana* and *Ziziphus spina-christi*. These bands arranged in an increasing order of water affinity. According to their age and status of water affinity, mayas are classified into three types: wet young mayas, moist mature mayas and dry old mayas (AbdelHameed et al., 1997).

The work of Abdel Hameed et al. (1997) on the watershed management and drainage systems of the Dinder river and its tributaries forms the baseline for mapping the drainage network of the DNP.

The drainage network of the park has been classified by Abdelhameed et al. (1997) into the Khor Gelagu network, Khor Masaweek drainage network, East-bank of the Dinder river network and West-bank of Dinder river network.

*Figure 2.9: The edges of mayas are surrounded by trees in a consistent pattern (Pictures taken by Khalid Hassaballah, March 2011).*

**Khor Gelagu drainage network**

Khor Gelagu is the main tributary of the Dinder river. The drainage network includes many mayas, the largest of which is Ras Amir (4.5 $km^2$) located 13 km northeast of the Gelagu camp. In older maps, it was referred to as Lake Ras Amir. It rarely dried up before 1970, and since then, it became less perennial and it is drying up haphazardly every few years.

**Khor Masaweek drainage network**

The Khor is also a large tributary of the Dinder river. Eleven prominent mayas can be found in this drainage network. Sambarok is the largest with an area of around 2.3 $km^2$.

**The Eastern bank of the Dinder river drainage network**

This includes the tributaries Ein El Shamis, Musa maya, Simseer and El Abyad.

**The Western bank of the Dinder river drainage network**

This includes Gererrisa (2 $km^2$) located 5 km north-west of the Gelagu camp, El Dabkara, Beit El Wahash (3.6 $km^2$) and Simaaya about 25 km south of the Gelagu camp.

## 2.2.5 Ecosystem of the Dinder National Park

The ecological water supply from the Dinder river supports the rich ecosystem of the DNP, both the aquatic species and wildlife in the park. The park supports a large population of wildlife during the dry season (Figure 2.10) and a smaller number during the rainy season. Dasmann (1972) classified the vegetation of DNP into four categories: wooded grassland; open grassland, woodland and riverine forest. While, the vegetation assessment by Hakim et al. (1979) and Abdel Hameed et al. (1996a) recognized three types of ecosystems, namely the *Acacia seyal-Balanites aegyptiaca* (Dahara) ecosystem, the riverine ecosystem and the mayas ecosystem (Figure 2.11). These ecosystems are composed of diverse communities with relatively few species. Uncontrolled mechanized farming and clearance of the tree cover have increased land degradation around the DNP.

*Figure 2.10: Wildlife diversity in the Dinder National Park during the dry season, (Pictures taken by Khalid Hassaballah, May 2011).*

*Figure 2.11: The three types of ecosystems in the DNP, a) the Acacia seyal-Balanites aegyptiaca (Dahara) ecosystem, b) the riverine ecosystem and c) the mayas ecosystem (Pictures taken by Khalid Hassaballah, a) in May 2011,b) and c) in November 2011).*

The Dinder river flows into the DNP where meeting the ecological water requirements is of vital importance to both the aquatic species and terrestrial animals in the park. It was noticed that some of the mayas do not receive water during recent years, while others which used to be dry were filled. The hydrology of filling/emptying of the mayas is not fully understood.

Floods play an important role for the hydrological and ecological integrity as a connectivity corridor between the rivers and the mayas floral and faunal habitat. The absence of floods leads to dryness of the mayas and loss of biodiversity (e.g. many aquatic and terrestrial plant species have disappeared and some are subjected to severe threats). This led to disappearance and migration of wildlife. Loss of biodiversity and degradation of mayas ecosystems have caused increasing concern about the current situation of the DNP, however, the causes are not yet fully understood. The park has an economic, environmental and social values, and provides a huge range of ecosystem services to the communities living within and around the park. The provided services are presented in Table 2.1. Therefore, conservation of the DNP ecosystems for direct and indirect human benefits is one of the major challenges facing Sudan.

*Table 2.1: The different ecosystem services provided by the DNP as categorized in the Millennium Ecosystem Assessment of 2005 (provisioning, regulating, supporting and cultural services).*

| ***Provisioning services*** | ***Regulating services*** | ***Supporting services*** | ***Cultural/Social services*** |
|---|---|---|---|
| ***Food:*** *Mayas Ecosystem provides the conditions for growing food for both human and wild animals. Mayas provide fish for human consumption and grass for wild animals. Forests also provide food for human consumption such as honey.* | ***Local climate and air quality:*** *Trees provide shade for wild animals whilst forests influence rainfall both locally and regionally. Trees or other plants also play an essential role in regulating air quality by removing pollutants from the atmosphere.* | ***Habitats for species:*** *The DNP ecosystems provide different habitats for many individual plant or animal that are essential for a species' lifecycle to survive. Migratory species including mammals, birds and fish are all depending on different ecosystems during their migrations.* | ***Tourism and education:*** *High potential opportunity for tourism and education (e.g. attractive place for local people and foreigners and opportunity for research and training). Thus, it provides considerable economic benefits and is a potential source of income for the country.* |
| ***Fresh water:*** *Mayas ecosystems play a vital role in the local hydrological cycle, as they regulate the flow. Vegetation and forests influence the quantity of water available locally and further downstream.* | ***Carbon sequestration and storage:*** *Ecosystems regulate the global climate by storing and sequestering greenhouse gases. In this way forest ecosystems in the DNP are carbon stores. Biodiversity also plays an important role by improving the capacity of ecosystems to adapt to the effects of climate change.* | ***Nutrient cycling:*** *Mayas ecosystems regulate the flows and concentrations of nutrients through a number of complex processes that allow these elements to be extracted from their mineral sources or recycled from dead organisms.* | ***Aesthetic appreciation and inspiration for culture and art:*** *The DNP Biodiversity and natural landscapes have been the source of inspiration for much of the art, folklore and culture in Sudan.* |

| ***Raw materials and medicinal resources:*** *The DNP ecosystems provide a great diversity of materials for construction and fuel including wood and charcoal. The DNP ecosystems also provide many plants used as traditional medicines for local people.* | ***Moderation of extreme events:*** *The DNP plays an important role in modulating the effects of extreme events. For example, prevent or reduce flooding. Maya wetlands attenuate floods by absorbing runoff peaks and storm surges.* | ***Maintenance of genetic diversity:*** *Some habitats have an exceptionally high number of species which makes them more genetically diverse than others.* | ***Economic benefits:*** *Tourism, Jobs for wildlife police, forestry, fishermen and honey collectors...etc.* |
|---|---|---|---|

# 3

# THE LONG-TERM TRENDS IN HYDRO-CLIMATOLOGY OF THE DINDER AND RAHAD BASINS

[3] This chapter is based on: Hassaballah K, Mohamed YA, Uhlenbrook S (2019) The long-term trends in hydro-climatology of the Dinder and Rahad basins, Blue Nile, Ethiopia/Sudan. International Journal of Hydrology Science and Technology 9:690-712 doi:10.1504/IJHST.2019.103447.

## SUMMARY

This chapter examines the long-term trends of streamflow, rainfall, and temperature over the Dinder and Rahad river basins. Streamflow of the Rahad river showed significant increasing trends in both the annual and seasonal flows. There was no detectable change in the mean annual and seasonal flow patterns of the Dinder. However, the analysis of seasonal maxima suggested a shift towards decreased flows during the high flow period (August) and increased flows during the low flow period (November). The Dinder maxima of August decreased from 517 $m^3/s$ over the early part of the record (1972-1991) to 396 $m^3/s$ over the latest years (1992-2011). The mean annual temperature showed significant increasing trends at the rate of 0.24 and 0.30 °C/decade in the examined stations. Rainfall showed no significant change. The result of this study suggests other factors than climate variability (e.g. land use and land cover changes) to be responsible for streamflow alterations.

## 3.1 INTRODUCTION

The headwater catchments of the Dinder and Rahad basins (D&R) shared between Ethiopia and Sudan generate over 7% of the Blue Nile basin's annual flow. The Rahad river supplies water to the Rahad irrigation scheme (126,000 ha), while the Dinder river is the main source of water for the diverse ecosystem of the Dinder National Park (DNP) in Sudan (see section 2.2.2). However, during recent years, the Dinder river has experienced significant changes in floodplain hydrology and water supply to local wetlands (mayas). These changes claimed to be due to climate and/or land use land cover changes. This has significant implications on the ecosystem functions and hence the services of the DNP (see section 2.2.3). Therefore, understanding the climate variability/change and its hydrological impacts is essential for water resources management, as well as for sustainable ecosystem conservation in the DNP.

Hydro-climatic variability plays a pivotal role in structuring the biophysical environment of riverine and floodplain ecosystems. Variability is natural but can also be enhanced by anthropogenic interventions. Alterations of hydro-climatic variables can have significant impacts on the ecohydrological functions of rivers and related ecosystems. Loss of biodiversity and degradation of ecosystems have caused increasing concern about the current situation of the D&R, particularly the ecosystems of the DNP in Sudan. However, the causes are not yet fully understood. Conservation of ecosystems for direct and indirect human benefit is one of the major global challenges.

Trend analysis for hydrological and meteorological time series is an important and common method for understanding climate variation and its impacts on water resources (Burn and Hag Elnur, 2002; Kahya and Kalaycı, 2004). The existence of a trend in hydrologic time series can be explained by the change in streamflow (e.g. Lins and Slack, 1999; Woo and Thorne, 2003; Cigizoglu et al., 2005). It can also be explained by changes in rainfall (e.g. Lettenmaier et al., 1994; Rodriguez-Puebla et al., 1998; Partal and Kahya, 2006; Shang et al., 2011). Temperature trends were analyzed to understand links to hydrology as a proxy for changes in evapotranspiration. (e.g. Ghil and Vautard, 1991; Stafford et al., 2000; Vinnikov and Grody, 2003; Mengistu et al., 2014).

For trend detection, nonparametric tests are more often used than the parametric ones. This was due to their suitability for data with specific distribution (e.g. non-Gaussian). The common nonparametric trend detection tests are; the Spearman's rho (SMR) (Spearman, 1904; Lehmann and D'abrera, 1975; Sneyers, 1990), and the Mann–Kendall (MK) test (Mann, 1945; Kendall, 1975). Close agreement with respect to the performance of these methods was found between MK and SMR (Yue et al., 2002) and MK and Cumulative Rank Difference (Onyutha, 2016). Accordingly, the MK test was applied in this study.

Many studies around the world used the Mann-Kendall (MK) test to identify hydro-climatologic trends. Tesemma et al. (2010), analyzing the trends of rainfall and streamflow over a 40-year period (1963-2003), showed no change of rainfall over the Blue Nile basin. Streamflow analysis for Bahir Dar and Kessie at the upper portion of the Blue Nile basin, and El Diem at the border between Sudan and Ethiopia showed that the annual streamflow did however indicate increased flow in the upper Blue Nile, but not at El Diem. Using MK and Pettitt tests, Gebremicael et al. (2013) found no significant change of the annual rainfall over the Upper Blue basin between the 1970s and the beginning of the 21st century. Nevertheless, both tests showed a statistically significant increasing trend of streamflow during the long rainy season (June-September) and the short rainy season (March-May), and a decreasing trend in the dry season (October-February) streamflow. The annual streamflow has increased significantly during the period (1971-2009). Since the Upper Blue Nile basin is neighboring the D&R, similarities of catchment characteristics could be expected, though differences cannot be excluded. Tekleab et al. (2013), studied the trends of rainfall, temperature, and streamflow within the Abay/Upper Blue Nile basin. The results showed statistically significant increasing and decreasing trends in the streamflow. Temperature showed increasing trends in most of the studied stations. In contrast, rainfall did not show any significant trends.

Recently, Masih et al. (2014), who reviewed droughts on the African continent, stated that the available evidence from the past clearly shows that the continent is likely to face extreme and widespread droughts in the future. They speculate that drought challenge is likely to aggravate because of slow progress in drought risk management, increasing population and demand for water, and degradation of land and environment. In contrast, Basheer et al. (2016), assessing the impacts of future climate change (2020s, 2050s, and

2080s) on the Dinder river flow and its possible implications on the DNP ecosystems, found that the climate will become warmer and wetter.

Thus, it has been shown that a variety of probable climatic impacts on the hydrologic system of the D&R are likely to happen. Therefore, the objective of this study is to investigate the long-term variations of streamflow, rainfall, and temperature over the D&R. The non-parametric Mann-Kendall (MK) and Pettitt tests (Pettitt, 1979) were applied to analyze the trends and to identify the changing points.

Streamflow regime is essential in sustaining ecological integrity of river systems (Poff et al., 1997). Therefore, The Indicators of Hydrologic Alterations (IHA) approach (Richter et al., 1996) was then applied to support the MK test and to analyze the essential characteristics of the streamflow likely to impact ecological functions in the D&R basin, including: flow magnitude, flow timing and rate of change in river flows. Understanding the level to which the streamflow has changed from its natural conditions is crucial for developing an effective management plan for ecosystem restoration and conservation.

In this study, the Pettitt test indicates that the changing points of streamflow in the Dinder and Rahad occurred during the late 1980s and the early 1990s. Therefore, to evaluate the current Dinder and Rahad rivers hydrology relative to historical conditions, the natural ranges of flows variations for both rivers have been characterized using the IHA approach for comparing two periods (1972-1991) and (1992-2011), hereafter defined as pre- and post-altering, respectively. We have made the subdivision from 1992 instead of 1990, to obtain an equal number of years before and after the changes for the IHA-based statistical comparison.

## 3.2 METHODS AND DATA USED

### 3.2.1 Trend detection tests

In this study, the non-parametric Mann-Kendall (MK) and Pettitt tests were applied to analyze the trends and the changing points of three hydro-climatic data time series of streamflow, rainfall, and temperature. Trends have been assessed in different time periods and varying lengths based on data availability. The MK statistic is given by:

$$S = \sum_{i=1}^{n-1} \sum_{j=i+1}^{n} \mathrm{Sgn}(X_j - X_i) \qquad (1)$$

Where $S$ is the MK statistic, $X_i$ and $X_j$ are the observations with $j > i$, $n$ is the time series data set length, and the sign function is given by:

$$Sgn(\theta) = \begin{bmatrix} +1 & \text{if} & \theta > 0 \\ 0 & \text{if} & \theta = 0 \\ -1 & \text{if} & \theta < 0 \end{bmatrix} \quad (2)$$

The variance Var(S) and the standard normal variate Z are calculated with Eqs. (3) and (4), respectively. The trend results in this study have been assessed at 5% significant level.

$$Var(s) = \frac{1}{18}\left[n(n-1)(2n+5) - \sum_{t} t_i(t_i-1)(2t_i+5)\right] \quad (3)$$

Where $t_i$ is the extent of any given tie, and $\Sigma_t$ denotes the summation over all ties. $H_0$ should be accepted if $|z| \leq z\alpha/2$ at the $\alpha$ level of significance.

$$Z = \begin{cases} \dfrac{s-1}{\sqrt{Var(s)}} & \text{if} \quad s > 0 \\ 0 & \text{if} \quad s = 0 \\ \dfrac{s+1}{\sqrt{Var(s)}} & \text{if} \quad s < 0 \end{cases} \quad (4)$$

The magnitude of the slope β, determined by (Hirsch et al. 1982) is given by:

$$\beta = \text{Median}\left[\frac{(X_j - X_i)}{(j-i)}\right] \quad \text{where } 1 < i < j < n \quad (5)$$

Where $X_i$ and $X_j$ are the data values at time *i* and *j*, respectively and *n* is the length of the whole data set.

The existence of increasing or decreasing trends was tested using the MK test. Then, the Pettitt test was applied to detect the changing points. The Pettitt test is a non-parametric test used to identify a single change-point in the data series if any (Pettitt, 1979). The significance of trends in the dataset is defined as “no significant trend”, “significantly increasing or decreasing trend” based on the defined confidence level of 5%. The MK computes Kendall's statistics (S), Kendall's tau (τ) and MK’s Z statistic. Positive Z values indicate increasing trends whereas negative values indicate decreasing trends. Finally, a probability (p-value) was computed and compared with the user-defined significance level in order to identify the trend of variables.

## 3.3 Indicators of Hydrologic Alterations (IHA)

The IHA technique is part of the Range of Variability Approach (RVA) developed by Richter et al. (1997). It is used to assess river ecosystem management goals defined based on a statistical representation of ecologically related hydrologic parameters (Richter et al., 1996). These parameters describe five essential characteristics of river flow that have ecological implication (Richter et al., 1996; Poff et al., 1997; Scott et al., 1997). The IHA technique computes 33 hydrologic parameters for each year.

For analyzing the alteration between two periods, the RVA described in Richter et al. (1997) was applied using the IHA software developed by The Nature Conservancy (2009). In RVA analysis, the pre-altering data for each parameter is divided into three categories. In this study, boundaries between categories were defined based on the default percentile values for non-parametric RVA analysis by adjusting the category boundaries 17 percentiles from the median. This ensures that in most conditions an equivalent number of values will fall into each category and gives three categories of equal size as given in Eq. (6):

$$C_l \leq P^{33} < C_m \leq P^{67} < C_h \tag{6}$$

Where, $C_l$, $C_m$ and $C_h$ are the low, middle and high categories, respectively. $P^{33}$ and $P^{67}$ are the 33rd and 67th percentiles, respectively.

The IHA software next computes the expected frequency with which the "post-altering" values of the IHA parameter should fall within each category, based on the pre-altering frequencies (in the non-parametric default, this would be 33% of the annual values in each of the three categories). Then it computes the frequency with which the "post-altering" annual values of IHA parameters actually fell within each of the three categories.

The Hydrologic Alteration (HA) factor is calculated for each of the three categories as given in Eq. (7):

$$HA = \frac{f_o - f_e}{f_e} \tag{7}$$

Where; $f_o$ is the observed frequency, and $f_e$ is the expected frequency.

Hydrologic Alteration with a positive deviation indicates an increasing in frequency of the value within the target category compared to the pre-altering period, while a negative deviation indicates a decreasing (The Nature Conservancy, 2009).

For assessing hydrologic alteration in the Dinder and Rahad rivers, the natural ranges of flows variations for both rivers have been characterized using the IHA based on variations in streamflow characteristics between two periods (1972-1991) and (1992-2011), hereafter defined as pre- and post-altering periods, respectively. Natural temporal variability of flow data was analyzed from Al-Gewisi station on the Dinder river and Al-Hawata station on the Rahad river.

To calculate the significant count for the deviation values, the IHA software randomly shuffles all years of input data and recalculates (fictitious) pre-altering and post-altering medians and coefficients of dispersions 1000 times. The significance count is the fraction of trials for which the deviation values of the medians or coefficients of dispersions were greater than of the real case. A low significance count (minimum value is 0) indicate that the difference between the pre-altering and post-altering periods is highly significant, and a high significance count (maximum value is 1) indicate that there is little difference between the pre-altering and post-altering periods. The significance count was interpreted similarly to a p-value in MK statistics.

## 3.4 HYDRO-CLIMATIC DATA

The hydro-climatic variables streamflow, rainfall, and temperature are selected because of a) the spatially assimilated hydrologic response that they provide, and b) they are the only variables having available long records of data. The temperature was used as a proxy for evapotranspiration. Table 3.1 shows the available data and their minimum, maximum and mean annual values.

There are twelve rainfall stations spatially distributed over the study area. Data are monthly. Six stations are in Rahad basin: Gedarif, Gadambaleya, Samsam, Um Seinat, Doka and Al-Hawata. Since there is no station with long records in the Dinder basin, data from four nearby stations were used: Ad Damazin, Abu Naama, Um Benien and Sennar. The same is true for the upper part of the catchment, so data from two nearby stations Gonder and Bahir Dar in the Ethiopian plateau with long records were used (Figure 3.1).

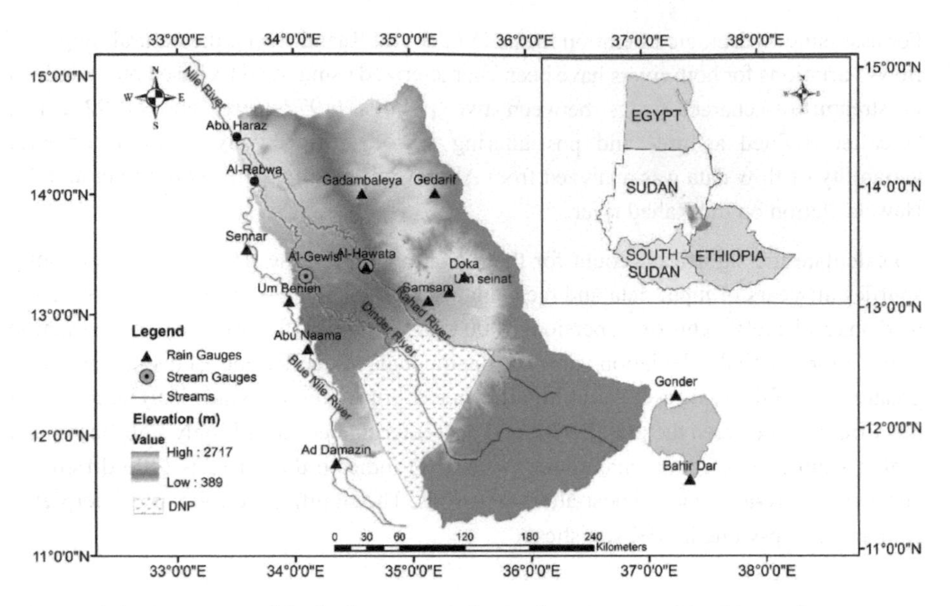


*Figure 3.1: Locations of the hydro-meteorological stations used in this study.*

Daily streamflow records for 40 years (1972-2011) at two hydrological stations (Al-Gewisi and Al-Hawata) on the Dinder and Rahad rivers respectively, were obtained from the Ministry of Irrigation and Water Resources, Sudan (MoIWR). The daily data were used for analyzing the streamflow parameters of magnitude, timing, and rate of change of flows using the IHA approach. The monthly mean, maximum, mean annual and maximum annual streamflow were calculated for the trend's analysis. At the downstream part of the D&R, only Gedarif station have a long record of temperature data. Gedarif was considered representative for this part of the basin (Figure 3.1). The monthly mean temperature records for the period (1941-2012) were obtained from the Sudanese Meteorology Authority. Since there is no temperature data in the upper part of the D&R basin, the neighboring station of Gonder was considered representative for the upper part of the basin (Figure 3.1). The temperature data for Gonder station close to the upper part of the D&R was obtained from the Ethiopian National Meteorological Agency. The hydro-climatic data in the D&R is generally scant with many data gaps and may contain measurements and/or typos errors. Abnormal values and outliers could lead to a wrong conclusion. Therefore, removing such errors is critical in data mining and data analysis, especially when analyzing trends. Thus, we carefully examined all the data before further statistical analysis. Data screening and data quality checks were performed for all data sets before analysis. Visual inspection and regression analysis between neighboring stations were used to identify outliers and fill in missing data in the data sets (if appropriate). For instance, we found that temperature data are accurate, while streamflow data contained outliers and typos errors, which were corrected as far as possible. Different methods have been used to fill in the missing gaps in rainfall and streamflow records. Regression analysis was used to fill in the missing gaps in monthly rainfall. Continuous missing data for a length of one year were omitted from the analysis. In the time series flow data, only missing data of a short duration (e.g. 1-3 days) was filled. Linear interpolations and rating curves were used to fill the short duration gaps in flow data. Missing flow data for a length of one month and above were omitted from the analysis.

Due to data scarcity in the region, some of the climate data were obtained from neighboring stations outside the case study boundary, but within the same climate zone. We analyzed the reference evapotranspiration ($ET_0$) in the region to support the analysis of using neighboring stations. Long-term monthly $ET_0$ data for some of the examined stations inside and outside the case study boundary was obtained from IWMI Online Climate Summary Service Portal (http://wcatlas.iwmi.org/results.asp). The analysis has shown that the $ET_0$ for the examined neighboring stations (i.e. Gonder, Bahir Dar, Ad Damazin, Abu Naama and Sennar) have similar patterns to those stations inside the case study boundary (i.e. Al-Hawata). Figure 3.2 shows the long-term monthly $ET_0$ for some of the examined stations in the region.

*Table 3.1: Available monthly rainfall data in the study area given as annual values (mm/a)*

| Station | Name | Data availability | $P_{max}$ | $P_{min}$ | $P_{mean}$ |
|---|---|---|---|---|---|
| 1 | Gedarif | 1903-2012 | 1035.3 | 289 | 608 |
| 2 | Gadambaleya | 1979-2012 | 779.6 | 303 | 544 |
| 3 | Samsam | 1979-2012 | 886 | 427 | 701 |
| 4 | Um Seinat | 1979-2012 | 1070 | 379 | 665 |
| 5 | Doka | 1979-2012 | 1016 | 414 | 682 |
| 6 | Al-Hawata | 1979-2012 | 917 | 222 | 511 |
| 7 | Sennar | 1907-2008 | 758 | 175 | 440 |
| 8 | Ad Damazin | 1981-2000 | 899 | 497 | 700 |
| 9 | Abu Naama | 1984-1998 | 815 | 372 | 606 |
| 10 | Um Benien | 1984-1998 | 715 | 313 | 507 |
| 11 | Gonder | 1953-2007 | 1823 | 720 | 1117 |
| 12 | Bahir Dar | 1961-2007 | 2036.7 | 894.5 | 1420 |

*Source*: *(Sudanese Meteorological Authority and Global Historical Climatology Network (GHCN)).*

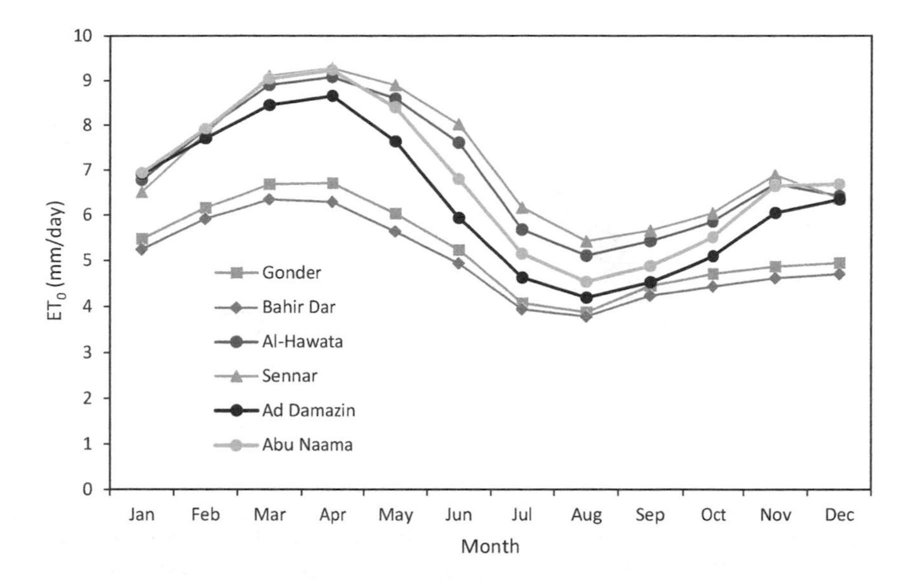


*Figure 3.2: Reference evapotranspiration ($ET_0$) for the examined stations (Gonder, Bahir Dar, Ad Damazin, Abu Naama and Sennar) outside the case study boundary, and (Al-Hawata) stations inside the case study boundary.*

The hydrological gauging stations are generally well maintained and discharge measurement using current meter is annually performed to update the rating curves in case of sedimentation or scouring. Figure (3.3) shows the rating curves at Al-Gewisi and Al-Hawata stations on the Dinder and Rahad rivers, respectively. The measured flow and gauge data used to derive the rating curves covered the period (1988-2012). The curves indicate downward shifts caused by morphological changes at the measurement sites as a result of sedimentation. The curves also confirm that the rating curves at both sites are regularly updated if appropriate. The Al-Hawata station located some 47 km upstream of Abu Rakham Barrage, and the station is about 22 km far from the possible effects of the back-water curve of the small reservoir created upstream of the Barrage."

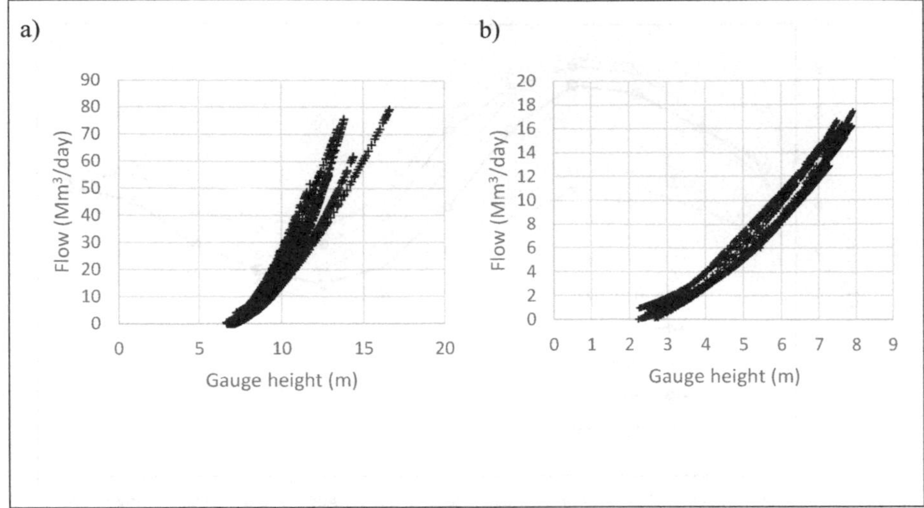


*Figure 3.3: Rating curves for the measured discharges at a) Al-Gewisi station in Dinder River and b) Al-Hawata station in the Rahad River for the period (1988-2012).*

## 3.5 Results and Discussion

This section presents the results of the statistical tests to assess long term trends of the D&R hydro-climatology. To verify the statistical test and to have critical analysis of streamflow, alteration in streamflow through IHA approach is also discussed.

### 3.5.1 MK and Pettitt analysis

***Trend of rainfall***

For the twelve gauging stations of rainfall (Table 3.2), the MK test shows no significant trends at 5% confidence level, over the D&R basin. Only one station (Doka) shows a significantly increasing trend with values of 0.332 and 0.006 for $\tau$ and p, respectively (Table 3.2). This result agrees with the literature on rainfall trends over the neighboring basin of the Blue Nile. For example, Tesemma et al. (2010) showed no change in rainfall in the Blue Nile basin during (1963-2003). Gebremicael et al. (2013), investigating trends in rainfall in the Blue Nile with records between the 1970s and the beginning of the 21$^{st}$ century, found no significant change in the annual rainfall in the upper Blue Nile basin. Tekleab et al. (2013), applied the statistical MK test to study the trends in rainfall, temperature, and streamflow in the Abay/Upper Blue Nile basin and found no significant trends in rainfall in all inspected stations.

Those studies reported no significant trends in rainfall across the Abay/Upper Blue Nile basin, which includes the upper D&R basin. The MK results were found to be sensitive to the time domain.

*Table 3.2: Man-Kendall results of annual rainfall at the 12 examined rainfall stations.*

| Station | Kendall's tau | S | P-value | Trend |
|---|---|---|---|---|
| Bahir Dar | -0.09898 | -107 | 0.3320 | No significant change |
| Gonder | -0.1176 | -156 | 0.02217 | No significant change |
| Samsam | -0.1979 | -111 | 0.1023 | No significant change |
| Um Sienat | 0.1658 | 93 | 0.1728 | No significant change |
| Doka | 0.3333 | 187 | **0.0051** | Significantly increasing |
| Hawata | 0.2141 | 120 | 0.0777 | No significant change |
| Gedarif | -0.1260 | -755 | 0.0515 | No significant change |
| Gadambalyia | 0.0607 | 34 | 0.6247 | No significant change |
| Damazin | 0.3158 | 60 | 0.0537 | No significant change |
| Abu Naama | 0.2762 | 29 | 0.1659 | No significant change |
| Um Benien | 0.2952 | 31 | 0.1370 | No significant change |
| Sennar | -0.0533 | -228 | 0.4513 | No significant change |

***Trend of temperature***

As expected, both Gonder and Gedarif show a significant increasing trend of temperature (Table 3.3), with mean annual temperature increasing at 0.30 °C/decade in Gondar at the Ethiopian highland, and at 0.24 °C/decade in Gedarif at the Sudan low-lying. It is expected that increased temperature, particularly during the dry season (November - June), may influence evapotranspiration from the mayas, leading to increased dryness. The results from this study are in agreement with previous climate change studies in both the upper and lower Blue Nile basin, which reported increasing trends of temperature (Elagib and Mansell, 2000; Elshamy et al., 2009; Elagib, 2010; Nawaz et al., 2010).

*Table 3.3: Mann-Kendall results for Gonder and Gedarif mean annual temperature*

| Station | Kendall's τ | S | P-value | Slope (°C/a) | Trend |
|---|---|---|---|---|---|
| Gonder | 0.3227 | 111 | 0.02118 | 0.030 | Significantly increasing |
| Gedarif | 0.2668 | 682 | 0.00081 | 0.024 | Significantly increasing |

*S : The (Kendall) S-statistic value*
*τ : The Kendall rank-correlation coefficient (τ )*
*p : The p-value (computed probability)*

***Trend of streamflow***

The mean annual flow showed a significant increasing trend for the Rahad river at Al-Hawata station, but not for the Dinder river at Al-Gewisi station. While the annual maximum flow of the Dinder showed a significant decreasing trend, it is not the case for Rahad (Table 3.4). The statistical tests of the seasonal time series of Rahad showed significant increasing trends of the monthly mean for July, August and November. While the monthly maximum flow showed a significant decreasing trend in August flow and increasing trend in November flow of the Dinder river at Al-Gewisi station, there was no evidence for significant trend for the Rahad river at Al-Hawata station (Table 3.4). Since August is the period of high flow in both the Dinder and the Rahad rivers, increasing flow in this period leads to inundation of floodplain including mayas, while decreasing flow leads to dryness of mayas. Figure 3.4 shows the Pettitt tests results for the abrupt changing points. Significant abrupt changes for August maximum flow (flood period) and for November maximum flow (recession period) in Dinder river during the late 1980s and the early 1990s were observed. The decreasing of river flood magnitude leads to decreasing or even absence of water flowing to the mayas causing many mayas to be subjected to dryness. It has been observed that during the late 1980s and early 1990s the areas of some mayas inside DNP have radically decreased due to the variation in river discharge and sediment deposition processes (AbdelHameed et al., 1997). Such mayas can no longer store enough water to satisfy the needs of the wildlife populations throughout the dry season. Detailed analysis of IHA for the environmental flow components (magnitude, frequency, timing and rate of change of flow) are discussed below. Since Dinder river supports the ecosystem of the DNP floodplain (mayas), our IHA analysis focused on the alterations of the high extreme flow parameters.

The importance of large floods (flows equal to or greater than the 10-year return period flood) is to inundate the Dinder river floodplain wetlands (mayas). Therefore, alterations in the magnitude, frequency, timing and rate of changes of the annual large flood peaks are likely to affect the production of native river-floodplain flora and fauna. The small flood pulse (flows equal to or greater than bankfull flows but less than the 10-year return period flood) inundates the maya wetlands to a shallower depth, with the result that forage and water for wildlife remained available for only a short period of time.

*Table 3.4: Mann-Kendall tests results of annual and seasonal flow for Dinder at Al-Gewisi and Rahad at Al-Hawata at 5% confidence level (P = 0.05)*

| | River | Kendall's τ | S | P-value | Trend |
|---|---|---|---|---|---|
| **Mean annual** | Dinder | -0.146 | -114 | 0.189 | No significant trend |
| | Rahad | 0.256 | 200 | **0.020** | Significantly increasing |
| **Annual maxima** | Dinder | -0.280 | -0.220 | **0.010** | Significantly decreasing |
| | Rahad | 0.163 | 127 | 0.142 | No significant trend |
| **August maxima** | Dinder | -0.338 | -0.264 | **0.002** | Significantly decreasing |
| | Rahad | 0.195 | 152 | 0.079 | No significant trend |
| **November maxima** | Dinder | 0.232 | 174 | **0.041** | Significantly increasing |
| | Rahad | 0.170 | 130 | 0.130 | No significant trend |
| **July mean** | Dinder | 0.153 | -120 | 0.165 | No significant trend |
| | Rahad | 0.232 | 181 | **0.036** | Significantly increasing |
| **August mean** | Dinder | -0.210 | -164 | 0.057 | No significant trend |
| | Rahad | 0.282 | 220 | **0.010** | Significantly increasing |
| **November mean** | Dinder | 0.315 | 237 | **0.005** | Significantly increasing |
| | Rahad | 0.255 | 194 | **0.024** | Significantly increasing |

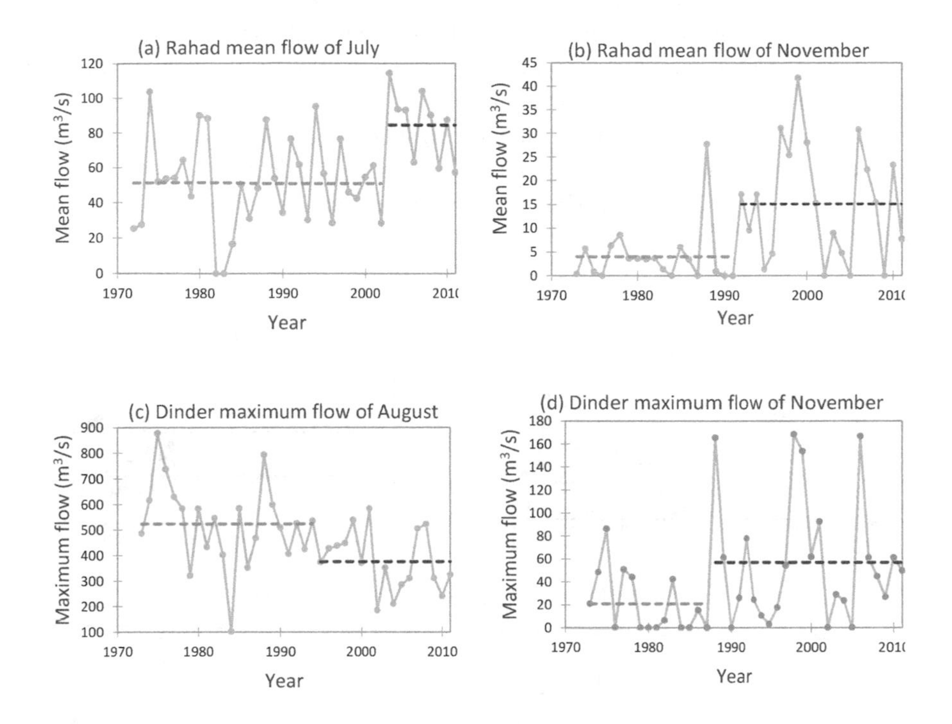


*Figure 3.4: The Pettitt homogeneity test for detecting the abrupt changing points of seasonal flows for (a) and (b) Rahad river, and (c) and (d) Dinder river. The dash lines are the mean of the time series before and after the change point.*

## 3.5.2 IHA analysis

***Magnitude of monthly flow***

The general pattern of median monthly flow of the Dinder river at Al-Gewisi station is that the median flow increased in July and November at the beginning and end of the rainy season (period of low flow) and decreased in August and October (period of high flow). The median monthly flows of July and November increased from 43 and 0 (dry) $m^3/s$ to 50 and 14 $m^3/s$, respectively. In contrast, the median monthly flows of August and October decreased by 20% and 11% from 266 and 101 $m^3/s$ to 210 and 90 $m^3/s$, respectively.

In comparison to Dinder and similarly to MK test, the Rahad median monthly flows showed increasing patterns in all months, with increasing pattern from 45, 133 and 0 $m^3/s$ to 65, 153 and 14 $m^3/s$ in July, August and November, respectively. The monthly flows are shown in (Figure 3.5 and 3.6). The alterations of the monthly flow magnitude between pre and post-altering periods in particular during months of high flows (August-October) are likely to affect habitat availability in particular on floodplains, which may lead to decrease or even disappearance of native plants species and increase in non-natives plants species that might not be suitable for the herbivores wildlife that inhabit the DNP.

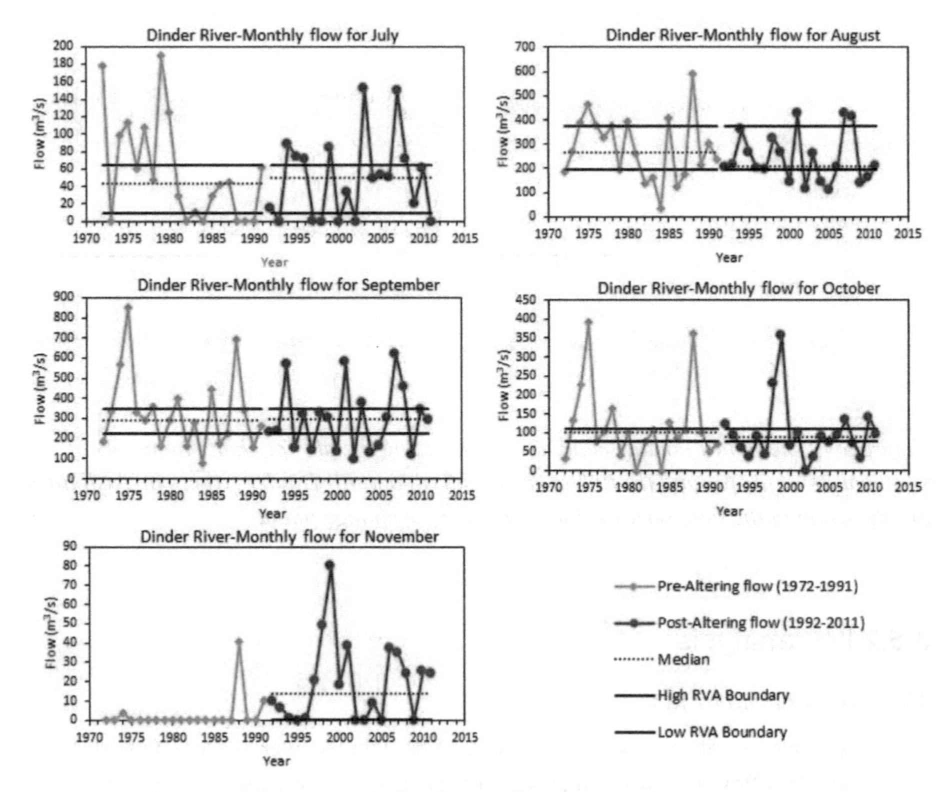


*Figure 3.5: Seasonal flow patterns for the Dinder river.*

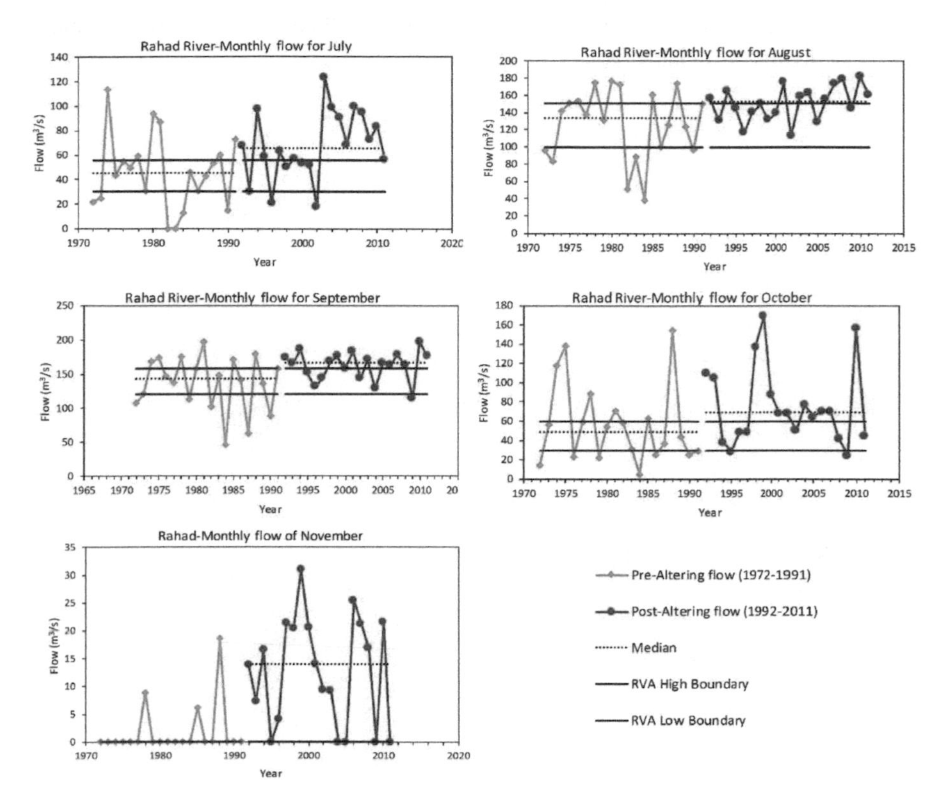


*Figure 3.6: Seasonal flow patterns for the Rahad river.*

### *Magnitude of river extreme floods*

Extreme floods are important in re-forming both the physical and biological structure of a river and its associated floodplains such as oxbow lakes and wetlands. For the maya wetlands of the DNP, all results show a decreasing maxima trend for the Dinder river. Figure 3.7 has shown that the post-altering median flow maxima for 1, 7, 30 and 90-day intervals in the Dinder river were, 14%, 13%, 15%, and 14%, lower than pre-altering. In contrast, in the Rahad river increasing patterns were observed, with post-altering median flow maxima for 1, 7, 30 and 90-day of 6%, 9%, 16%, and 21%, respectively, higher than pre-altering. Peak flows are critical aspects of the lateral connectivity between Dinder and Rahad rivers and its associated floodplains (mayas). The alterations in the Dinder river flow are likely to affect the ecosystems in DNP negatively. The decrease in magnitude of the annual flood peaks that reduce the amount of water flowing to the river-floodplain may reduce the production of native flora and fauna, and animal migration that may be linked to floodplain inundation.

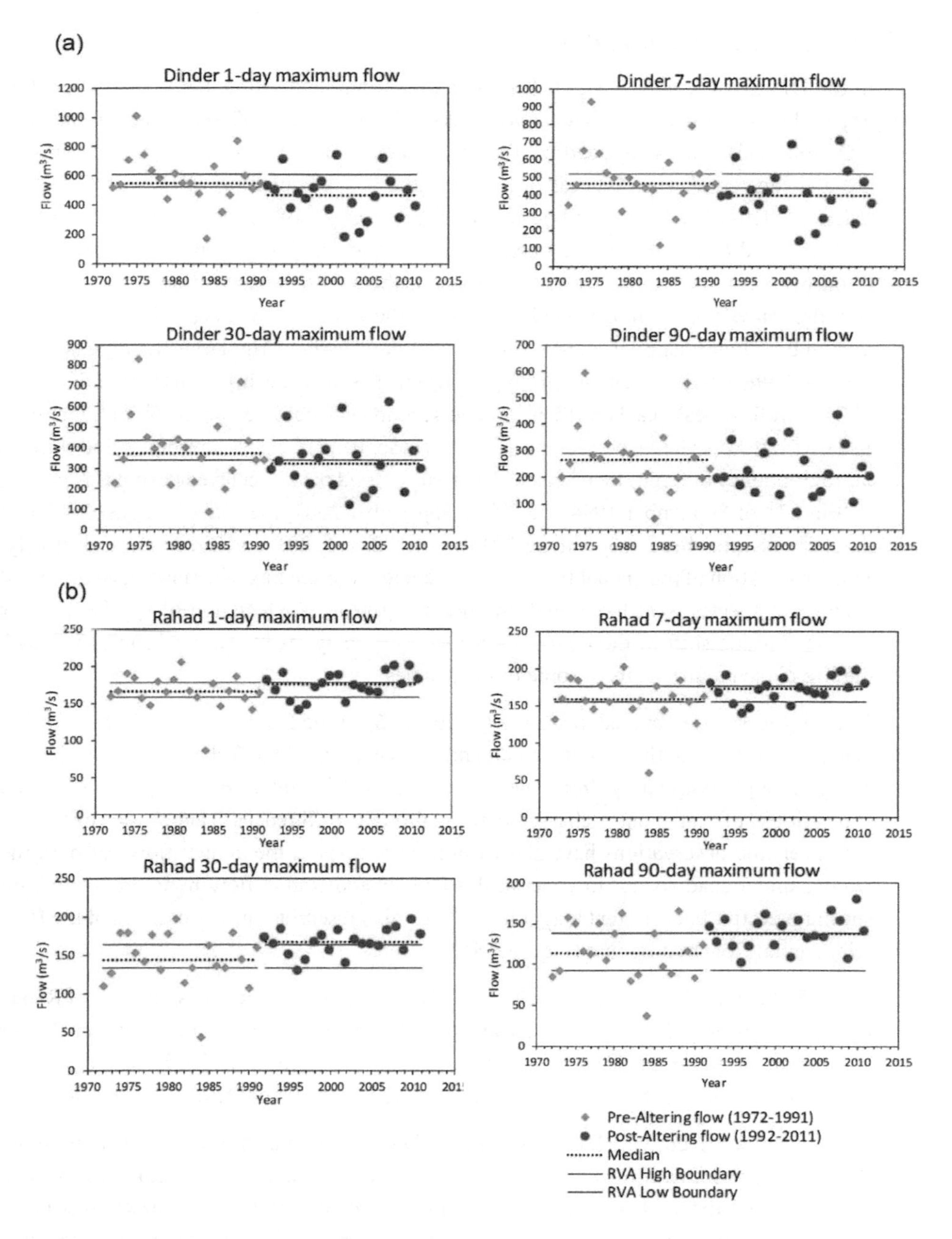


*Figure 3.7: Annual maximum flows of 1, 7 and 30-day for (a) Dinder river and (b) Rahad river.*

***Timing of annual extreme floods***

In the Dinder river, the timing of the annual maximum daily flow before and after flow impact happened within the same two weeks (16 September – 02 September, Julian date (JD) 260–246), but 14 days earlier. The large flood peak flows occurred twice during the pre-altering period. The first peak occurred on the 16th of September 1975, with flow peak reached 1010 $m^3/s$. The second peak occurred on the 2nd of September 1988, with flow peak reached 834 $m^3/s$. On the other hand, the post-altering period showed zero large flood peaks. In the Rahad river, the timing of the annual maximum daily flows before and after flow alteration happened also within the same two weeks (22 September – 10 September, Julian date (JD) 266–254), but 12 days earlier. The large flood peak flows occurred twice during the pre-altering period. The first peak occurred on the 9th of October 1974 with flow peak reached 190 $m^3/s$. The second peak occurred on the 5th of September 1981 with flow peak reached 206 $m^3/s$. On the other hand, the post-altering period showed more frequency of large flood peaks. Four large flood peaks occurred during this period; on the 17th of September 1994, the 2nd of September 2007, the 27th of August 2008 and the 20th of September 2010, with peak flows of 192, 196, 201, and 201 $m^3/s$, respectively. Synchronization of the annual flood with lifecycle requirements of a range of riverine and floodplain species is of likely high importance given the adaptation of species to their habitat. Timing shift of the Dinder river peak flow may lead to desynchronization with the life cycle requirements of some of the species.

The long-term mean annual discharge (1900-2016) of the Dinder river ($2.70 \times 10^9$ $m^3/a$) is about two to three times of the mean annual discharge of the Rahad river ($1.102 \times 10^9$ $m^3/a$), though originating from the same region. Therefore, one may expect some similarities of catchment characteristics, though differences cannot be excluded. However, the observations have shown high variations in the annual flows of both the Dinder and Rahad rivers. In general, the Dinder and Rahad flow hydrographs exhibit similar patterns, but different magnitudes. They also resemble similar patterns of the Blue Nile annual flow hydrograph (Figure 3.8).

In 1975, the annual flow of the Dinder is $5 \times 10^9$ $m^3/a$, four to five times the flow of Rahad in the same year ($1.3 \times 10^9$ $m^3/a$), as shown in Figure 3.8. The first flow peak in Dinder records occurred on the 16th of September 1975 is 1010 $m^3/s$, which is also about five times the flow peak of the Rahad river that occurred on the 9th of October 1974 (190 $m^3/s$).

The 1975 flood event which happened in the Dinder river basin is also happening in the Rahad river basin with flow magnitude of $1.3 \times 10^9$ $m^3/a$ compared to $1.4 \times 10^9$ $m^3/a$ in 1974. These indicate that the extreme events which happened in the Dinder river basin are likely to happen in the Rahad river, but not necessarily happening with the same magnitude and timing. This is likely due to temporal and spatial variability of rainfall and complexity of the runoff process over the upper Dinder and Rahad basins.

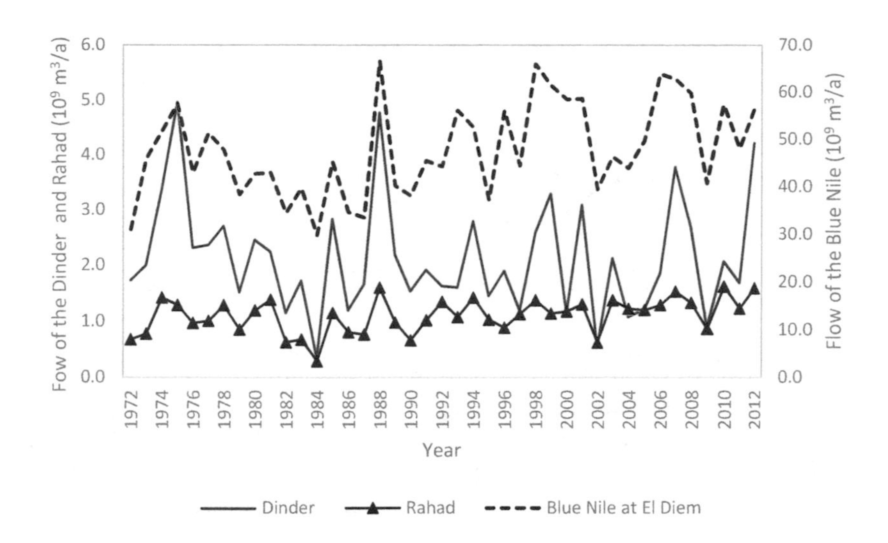


*Figure 3.8: Comparison of annual flow of the Dinder, Rahad and the Blue Nile (1972-2012).*

### *Rate of change in flow*

The median rate of flow rises (positive differences between consecutive daily values) in the Dinder river has decreased by 38% from 32 m$^3$/s/day during the pre-altering period to 20 m$^3$/s/day during the post-altering period. The median rate of flow falls (negative differences between consecutive daily values) has decreased by 53% from 17 m$^3$/s/day during the pre-altering period to 8 m$^3$/s/day during the post-altering period. Similar to Dinder, the median rate of flow rises in Rahad river has decreased by 40% from 5 m$^3$/s/day during the pre-altering period to 3 m$^3$/s/day during the post-altering period. The median rate of flow falls has decreased by 60% from 5 m$^3$/s/day during the pre-altering period to 2 m$^3$/s/day during the post-altering period (Figure 3.9). The rate of change in flow can affect persistence and lifetime for both aquatic and riparian species (Poff et al., 1997), particularity in such arid area where streamflow can change rapidly in a very short period of time due to excessive rainfall.

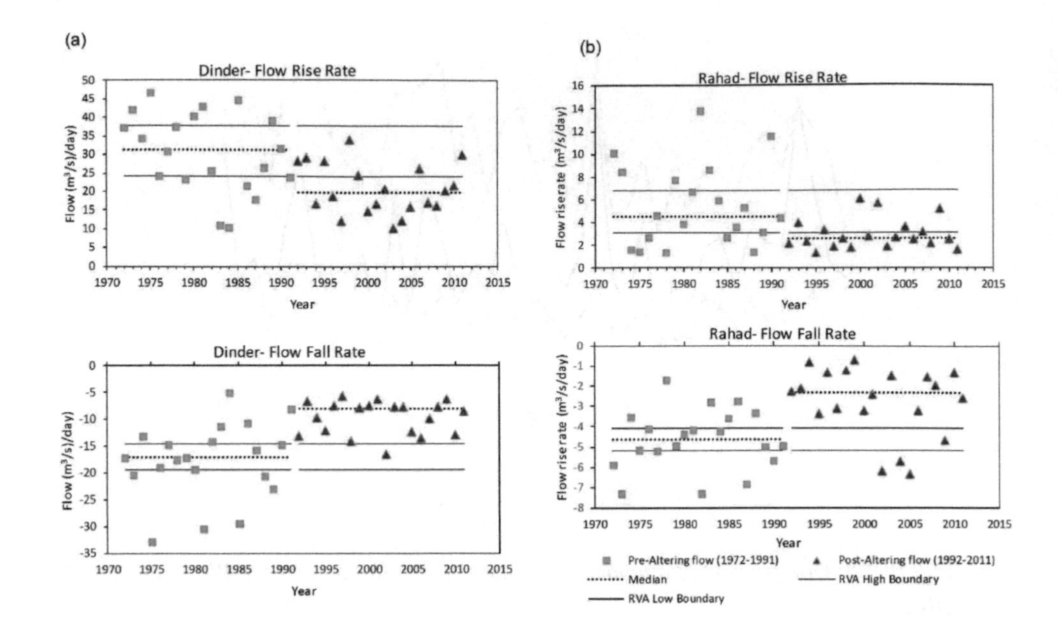


*Figure 3.9: Rates of flow rises and falls for (a) Dinder river and (b) Rahad river.*

## 3.6 Conclusions

The long-term trends of the Dinder and Rahad hydro-climatology have been analyzed for twelve rainfalls, two temperatures and two streamflow gauging stations, over different periods of time. The mean annual temperature showed statistically significant increasing trends at the rate of about 0.24 and 0.30 °C/decade in Gedarif and Gonder stations, respectively. No significant changes in rainfall have been detected. The trend results on rainfall agree with the literature on neighboring catchments of the Blue Nile (e.g. Tesemma et al., 2010; Gebremicael et al., 2013; Tekleab et al., 2013).

The mean annual streamflow of the Rahad river exhibited a statistically significant increasing trend, but not for the Dinder river which showed no significant changes. The trend of the monthly mean flows showed significant increasing trends in Rahad river for July, August and November, while no significant trend was observed in Dinder river. The monthly maxima flow showed a significantly decreasing trend of August maxima flows and decreasing trend of November maxima flows in the Dinder river, while no evidence for a significant trend of monthly maxima flows of the Rahad river. Reduction of the Dinder peak flow can have a direct impact on filling of the maya wetlands, the main water source for the DNP during the dry months. The Pettitt test indicates that the changing points of streamflow in the Dinder and Rahad occurred during the late 1980s and the early 1990s. The results of increasing temperature associated with increasing flow in Rahad river, indicate that the increasing trend of temperature shall not always lead to decreasing discharge as land use land cover change is another important factor in the partitioning of rainfall.

The IHA analysis has shown that the flow of the Rahad river was associated with significant upward alterations in some of the hydrological indicators. The flow of the Dinder river was associated with significant downward alterations. Particularly, these were:

a) a decrease in the magnitude of the river flow during August (peak flow) and an increase in low flows (November);

b) a decrease in magnitude of flow extremes (i.e. 1, 7, 30 and 90-day maxima); and

c) a decrease in flow rise rate and an increase in flow fall rate.

These alterations in the Dinder river flows are likely to affect the ecosystems in DNP negatively. The importance of the annual flood that inundates the Dinder river floodplain wetlands (mayas) is likely to have significant effects on a range of species that depend on the seasonal patterns of flow. Therefore, alterations in the magnitude of the annual flood that reduce the amount of water flowing to the mayas may reduce the production of native river-floodplain flora and fauna, and lead to migration of animals that are linked to wetlands inundation.

The result of no significant trends of rainfall over D&R coupled with increasing/decreasing trends of streamflow indicates other factors than climate variability (e.g. land use land cover changes) might be responsible for streamflow alterations. For this reason, analysis of streamflow response to land use and land cover changes has been examined and is presented in the next chapter.

# 4

# ANALYSIS OF STREAMFLOW RESPONSE TO LAND USE AND LAND COVER CHANGES USING SATELLITE DATA AND HYDROLOGICAL MODELLING

---

[4] This chapter is based on: Hassaballah, K., Mohamed, Y., Uhlenbrook, S., and Biro, K.: Analysis of streamflow response to land use and land cover changes using satellite data and hydrological modelling: case study of Dinder and Rahad tributaries of the Blue Nile (Ethiopia–Sudan), Hydrol. Earth Syst. Sci., 21, 5217-5242, https://doi.org/10.5194/hess-21-5217-2017, 2017.

## SUMMARY

Understanding the land use and land cover (LULC) changes and their implication on surface hydrology of the Dinder and Rahad (D&R) basins (77,504 $km^2$) is vital for the management and utilization of water resources in the basins. Although there are many studies on LULC changes in the Blue Nile basin, specific studies on LULC changes in the D&R basins are still missing. Hence, its impact on streamflow is unknown. This chapter aims to understand the LULC changes in the Dinder and Rahad and its implications on streamflow response using satellite data and hydrological modelling. The hydrological model has been derived by different sets of land use and land cover maps from 1972, 1986, 1998 and 2011. Catchment topography, land cover and soil maps are derived from satellite images and serve to estimate model parameters. The results of the LULC changes detection between 1972 and 2011 indicate a significant decrease in woodland and an increase in cropland. Woodland decreased from 42% to 14% and from 35% to 14% for Dinder and Rahad, respectively. Cropland increased from 14% to 47% and from 18% to 68% in Dinder and Rahad, respectively. The model results indicate that streamflow is affected by the LULC changes in both the Dinder and the Rahad rivers. The effect of LULC changes on streamflow is significant during 1986 and 2011. This could be attributed to the severe drought during the mid-1980s and the recent large expansion in cropland.

## 4.1 INTRODUCTION

Streamflow is an important hydrological variable needed for water resource planning and management and for ecosystem conservations. The rainfall runoff process over the upper Dinder and Rahad basins (D&R) is complex and non-linear and exhibits temporal and spatial variability (Hassaballah et al., 2016). To manage water resources effectively at a local level, decision makers need to understand how human activities and climate change may impact local streamflow. Therefore, it is necessary to understand the hydrological processes in the runoff-generated catchments and the possible interlinkages of the LULC changes with catchment runoff. For this reason, satellite data and hydrological modelling were used to analyze the LULC changes and their impacts on streamflow response in the D&R.

The D&R generate around 7% of the Blue Nile Basin's annual flow. The Rahad river supplies water to the Rahad irrigation scheme (126,000 ha), while the Dinder river supplies water to the diverse ecosystem of the Dinder National Park (DNP). The DNP (10,291 $km^2$) is a vital ecological area in the arid and semi-arid Sudanese–Saharan region.

The Dinder and Rahad rivers have experienced significant changes in floodplain hydrology in recent years, which some claim is caused by LULC changes in the upstream catchment. The floodplain hydrology defines the seasonal wetlands ("mayas") which are the only source of water in the DNP during the dry season (8 months). The hydrology of the mayas has large implications on the ecosystem of the DNP. A detailed description of the maya wetlands is provided in section 2.2.2.

LULC changes was identified as a key research priority with multi-directional impacts on both human and natural systems (Turner et al., 2007). Many studies highlighted the impacts of LULC changes on hydrology (e.g. DeFries and Eshleman, 2004; Uhlenbrook, 2007; Warburton et al., 2012), on ecosystem services (e.g. DeFries and Bounoua, 2004; Metzger et al., 2006; Polasky et al., 2011) and on biodiversity (e.g. Hansen et al., 2004; Hemmavanh et al., 2010). LULC changes is a widespread observable phenomenon in the Ethiopian highlands, as shown by Zeleke and Hurni (2001), Bewket and Sterk (2005), Hurni et al. (2005) and Teferi et al. (2013). These studies have pointed out different types and rates of LULC changes in different parts of the Ethiopian highlands over different time periods and reported that the expansion of croplands associated with a decrease in woodlands has been the general form of transitions.

Recently, Gumindoga et al. (2014) assessed the effect of land cover changes on streamflow in the upper Gilgel Abay river basin in northwestern Ethiopia. Their results showed significant land cover changes where cropland has changed from 30% of the catchment in 1973 to 40% in 1986 and 62% in 2001. The study attributed these changes to the increase in population, which increased the demands for agricultural land. The study has also shown that farmers in the area are commonly clearing forests to create croplands, and the resulting effect was the decrease in forested land from 52% in 1973 to 33% in 1986 and 17% in 2001. Since the upper Blue Nile basin is neighboring the D&R, one may expect some similarities of catchment characteristics, though differences cannot be excluded. These transitions have contributed to the high rate of soil erosion and land degradation in the Ethiopian plateau (Bewket and Teferi, 2009). Understanding the impacts of LULC changes on hydrology and incorporating this understanding into the emerging focus on LULC changes science are the most important needs for the future (Turner et al., 2003).

Many models have been developed to simulate impacts of LULC changes on streamflow. These can be categorized as an empirical black-box, conceptual, and physically based distributed models. Each type of these three models has its own advantages and limitations. Several situations in practice demand the use of simple tools such as the linear system models or black-box models. Nevertheless, these simpler models usually fail to mimic the non-linear dynamics, which are essential in the rainfall-runoff transformation process. Therefore, the development of a dynamic modelling language within a GIS framework such as PCRaster is a further important stage that allows complex models, such as the WFlow rainfall–runoff model, to be implemented, making use of globally available spatial datasets. The PCRaster programming language is an environmental modelling language to build dynamic spatial environmental models (Bates and De Roo, 2000; Karssenberg, 2002; Uhlenbrook et al., 2004). Such spatially distributed models also have the potential to help in answering questions of policymakers about the impact of spatial changes (e.g. impacts of LULC changes on streamflow dynamic). It has been shown that a variety of probable LULC changes impacts on hydrologic processes in the D&R are likely to happen. Therefore, the objective of this study is to understand the LULC changes in the D&R and its impacts on streamflow response using satellite data, GIS and remote sensing, as well as hydrological modelling. The WFlow distributed hydrological model (Schellekens, 2011) is used to simulate the processes. In addition, understanding the level to which the streamflow has altered is critical for developing an effective management plan for ecosystem restoration and conservation. Thus, the indicators of hydrological alteration (IHA) approach proposed by Richter et al. (1996) was then applied to analyze the streamflow characteristics likely to affect the ecological processes in the D&R, including flow magnitude, timing and rate of change of flow.

## 4.2 Data and Methods

Limited data are available for simulating the hydrology of the D&R. To fill this data gap, use has been made of globally available free datasets. The datasets which have been used to run the WFlow model are divided into two datasets: static data and dynamic data.

### 4.2.1 Input data

***Static data***

The static data contain maps that do not change over time. They include maps of the catchment delineation, Digital Elevation Model (DEM), gauging points, land use, local drainage direction (ldd), outlets and rivers. These maps were created with pre-prepared processes of the WFlow hydrologic model.

The catchment boundary has been delineated based on a 90 m x 90 m DEM of the NASA Shuttle Radar Topographic Mission (SRTM) obtained from the Consortium for Spatial Information (CGIAR_CSI) website (http://srtm.csi.cgiar.org).

Multi-temporal Landsat data for the years 1972, 1986, 1998 and 2011 were obtained free of charge from the internet site of the United States Geological Survey (USGS) (source: http://glovis.usgs.gov/). All images were geometrically corrected into the Universal Transverse Mercator (UTM) coordinate system (Zone-36N).

The soil map was obtained free of charge from the Food and Agriculture Organization (FAO) Harmonized World Soil Database (HWSD). The original catchment boundary layer provided 44 soil mapping unit (SMU) classes. These classes have been reclassified into 8 dominant soil group (DSG) categories, based on the DSG of each soil mapping unit code. This was necessary to reduce the model complexity. The WFlow soil model requires estimates of 8 parameters per soil type, which means 352 parameters if it is for 44 soil types. Therefore, reclassification of soil map into 8 dominant soil groups reduces the number of estimated parameters to 64. The categories are: vertisols (71%), luvisols (9%), nitisols (8%), leptosols (5%), cambisols (4%), alisols (2%) and fluvisols (1%). The map was then projected to WGS-84-UTM -zone-36N and resampled to a horizontal resolution of 500 m.

***Satellite based rainfall and evapotranspiration data***

The dynamic data contain maps that change over time. It includes daily maps of the precipitation and evapotranspiration. These maps were created with a pre-preparation step1 and step2 of WFlow model. In this study, three open-access satellites-based rainfall estimates (SBRE) products were compared based on their runoff performance at Al-Gewisi and Al-Hawata stations the outlets of the Dinder and Rahad basins, respectively. The best product was then used to run the WFlow model using different land use and land cover (LULC) maps. The SBRE and the evapotranspiration products used in this study are Rainfall Estimates (RFE 2.0), potential evapotranspiration (PET), Tropical Rainfall Measuring Mission (TRMM) and Climate Hazards Group InfraRed Precipitation with Stations (CHIRPS).

The RFE 2.0 and the PET data were obtained from the Famine Early Warning System Network (FEWS NET). The horizontal resolution is 0.1 degree (11.0 km) for the RFE and 1.0 degree (110 km) for PET. This data is available on a daily basis from 2001 to near real-time period of record. More details can be found at http://earlywarning.usgs.gov/adds/downloads/.

The TRMM is a joint space mission between NASA and the Japan Aerospace Exploration Agency (JAXA) launched in 1997. The TRMM satellite rainfall measuring instruments include the Precipitation Radar (PR), TRMM Microwave Image (TMI), a nine-channel passive microwave radiometer, a Visible and Infrared Scanner (VIRS), and five-channel visible/infrared radiometer (Huffman and Bolvin 2013). In this study, TRMM 3B42v7 which has a spatial resolution of 0.25° and a temporal resolution of 3 hours has been used. More information can be found at https://trmm.gsfc.nasa.gov/.

The CHIRPS data were developed by the Climate Hazards Group (CHG) and scientists at the U.S. Geological Survey Earth Resources Observation and Science Center. This product is a new quasi-global precipitation with daily to seasonal time scales, a 0.05° resolution, and 1981 to near real-time period of record. The CHIRPS data uses the monthly Climate Hazards Precipitation Climatology (CHPClim), the InfraRed (IR) sensors from the Group on Earth Observations (GEO) satellites, the TRMM 3B42 product, and the ground precipitation observations. More information about CHIRPS data can be found in Funk et al. (2014). A summary of all precipitation and evapotranspiration satellite products was provided in Table 4.1. All maps were projected into WGS-84-UTM-zone 36N (meters), clipped to catchment extent, and then resampled to a resolution of 500 m.

*Table 4.1: Summary of the different precipitation and evapotranspiration satellite products*

| Product | Developer | Spatial resolution | Covering area | Temporal resolution | Time span | Ground measurement |
|---|---|---|---|---|---|---|
| TRMM 3B42v7 | NASA, JAXA | 0.25° | 0°E-360°E/50°N-50°S | 3 hourly | Jan 1998 - present | Yes |
| RFE 2.0 | NOAA (CPC) | 0.1° | 20°E-55°E/40°N-40°S | 6 hourly | Jan 2001 - present | Yes |
| CHIRPS v2.0 | CHG | 0.05° | 0°E-360°E/50°N-50°S | daily | Jan 1981 - present | Yes |
| PET | NOAA (CPC) | 1.0° | 20°E-55°E/40°N-40°S | 6 hourly | Jan 2001 - present | Yes |

***Observed hydrological streamflow***

Daily streamflow data at Al-Gewisi station on the Dinder river and at Al-Hawata station on the Rahad river for the period (2001-2012) were obtained from the Ministry of Irrigation and Water Resources, Sudan. This data is mainly used for calibration and validation of the WFlow hydrological model.

## 4.3 LULC CLASSIFICATION AND CHANGE DETECTION

LULC images were selected in the same season to minimize the influence of seasonal variations on the classification result. All acquired images had less than 10% cloud cover. However, in order to cover the entire study area, more than eight scenes of the satellite data were processed (Table 4.2). Subsequently, all images were mosaicked and resampled to a pixel size of 30m × 30m. The classification results of the historical images 1972, 1986 and 1998 were validated through visual interpretation of the unclassified satellite images and supported by in-depth interviews with local elders. The classification of the 2011 image was validated by ground survey during a field visits throughout the study area during the period between 2011 and 2013, assuming no significant change during this period. A Global Positioning System (GPS) device was used to obtain exact location point data for each LULC class included in the classification scheme and for the creation of training sites and for signature generations as well. Moreover, field notes, site descriptions and terrestrial photographs were taken to relate the site location to scene features. A total of 120 training areas were selected based on image interpretation keys, established during the field survey and from interviews with the local people. This later step was used as a crosscheck validation for the visual interpretation performed to the historical images. A supervised maximum likelihood classification (MLC) technique was independently employed to the individual images. MLC is the most common supervised classification method used with remote sensing image data (Pradhan and Suleiman, 2009; Ellis et al., 2010). The derivation of MLC is generally acceptable for remote sensing applications and is used widely (Richards et al., 2006).

The accuracy assessment of the classified images was based on the visual interpretation of the unclassified satellite images (Biro et al., 2013). However, the visual interpretation was conducted by an independent analyst not involved in the classification. The stratified random sampling design, where the number of points was stratified to the LULC types, was adopted in order to reduce bias (Mundia and Aniya, 2006). Accordingly, error matrices as cross-tabulations of the classified data vs. the reference data were used to evaluate the classification accuracy. The overall accuracy, the user's and producer's accuracies, and the Kappa statistic values were then derived from the error matrices.

Multi-date post-classification comparison (PCC) change detection method described by Yuan et al. (2005) was used to determine the LULC changes in three intervals: 1972–1986, 1986– 1998 and 1998–2011. PCC is a quantitative technique that involves an independent classification of separate images from different dates for the same geographic location, followed by a comparison of the corresponding pixels (thematic labels) in order to identify and quantify areas of change (Jensen, 2005; Al Fugara et al., 2009). It is the most commonly used method of LULC changes detection mapping (Kamusoko and Aniya, 2009).

*Table 4.2: Description of used satellite images.*

| Acquisition date | Satellite | Number of scenes | Spectral bands | Spatial resolution |
|---|---|---|---|---|
| 04 Nov. & 11 Dec. 1972 | Landsat MSS | 9 | 1 – 4 bands | 60 m |
| 12 Nov. & 26 Nov. 1986 | Landsat TM | 9 | 1 – 6 bands | 30 m |
| 27 Nov. & 13 Dec. 1998 | Landsat TM | 8 | 1 – 6 bands | 30 m |
| 07 Nov. & 10 Dec. 2011 | Landsat TM | 8 | 1 – 6 bands | 30 m |

*MSS, multispectral scanner; TM, thematic mapper*

## 4.4 DESCRIPTION OF THE WFLOW HYDROLOGICAL MODEL

In order to assess the impacts of LULC changes on the stream-flow dynamic, the WFlow distributed hydrological model (Schellekens, 2011) is forced using SBRE. The WFlow is a state-of-the-art open-source distributed catchment model. The model is part of the Deltares OpenStreams project (http://www.openstreams.nl). The model is derived from the CQFLOW model (Köhler et al., 2006). It is a hydrological model platform that includes two models: the WFlow_sbm model described by Vertessy and Elsenbeer (1999), derived from the TIOPG_SBM soil concept, and the WFlow_hbv model (distributed version of the HBV model). The model directly appeals to the need within the hydrological and geomorphologic sciences community to effectively use spatial datasets, e.g. digital elevation models, land use maps, dynamic satellite data for rapid and adequate modelling of river basins with limited data availability. The model is programmed in PCRaster GIS dynamic language (Deursen, 1995).
In this study, the WFlow_sbm PCRaster-based distributed hydrological model, which makes use of the Gash and the TOPOG_SBM models, was used. The model requires less calibration and maximizes the use of available spatial data that make it a suitable model for this study. Step one of WFlow model was to delineate the river network and the gauging points based on the DEM. Next, a land use and soil maps were added to the model, and parameters were estimated based on physical characteristics of the soil and land use type. The rainfall interception was calculated using the Gash model (Gash, 1979; Gash et al., 1995), while hydrologic processes that cause a runoff or overland flow were calculated using the TOPOG_SBM model. The WFlow uses potential evapotranspiration as input data and derives the actual evaporation based on soil water content and vegetation cover type. The analytical model of rainfall interception in the WFlow is based on Rutter's numerical model (Gash, 1979; Gash et al., 1995). The surface runoff is modelled using a kinematic wave routine. Combination of the total rainfall and evaporation under saturated-canopy conditions is done for each rainfall storm to determine average values of precipitation and evaporation from the wet canopy. In case the soil surface is partially saturated, the rainfall that falls on the saturated area is directly added to the surface runoff component. The soil is represented by a simple bucket model that assumes an exponential decay of the saturated conductivity with depth. Lateral subsurface flow is simulated using the Darcy equation. Soil depth is identified for different land use types and consequently scaled using the topographic wetness index. Different parameters are assigned to each land cover type. These parameters include rooting depth, leaf area index (LAI), ratio of evaporation from wet canopy to average rainfall ($E_w/R$), albedo, canopy gap fraction and maximum canopy storage. All model parameters are linked to the Wflow model through lookup tables. The lookup tables are used by the model to create input parameter maps. Each table consists of four columns. The first column is used to identify the land use class,

the second column indicates the sub catchment, the third column represents the soil type and the last column lists the assigned values based on the first three columns. The parameters are linked to land use, soil type or sub-catchment through lookup tables. A description of the Wflow model parameters is presented in Appendix B and the calibrated values for each parameter are presented in Appendix C. The WFlow_sbm interception and soil model's equations are presented in Appendix A. Further details of the Wflow model are also given at https://media.readthedocs.org/pdf/wflow/latest/wflow.pdf. The model is fully distributed, which means that it makes the calculations for every grid cell of the basin. Each cell (500m×500 m) is seen as a bucket with a total depth divided into saturated and unsaturated stores (Figure 4.1). The streamflow model results were then analyzed using the IHA approach described by Richter et al. (1996).

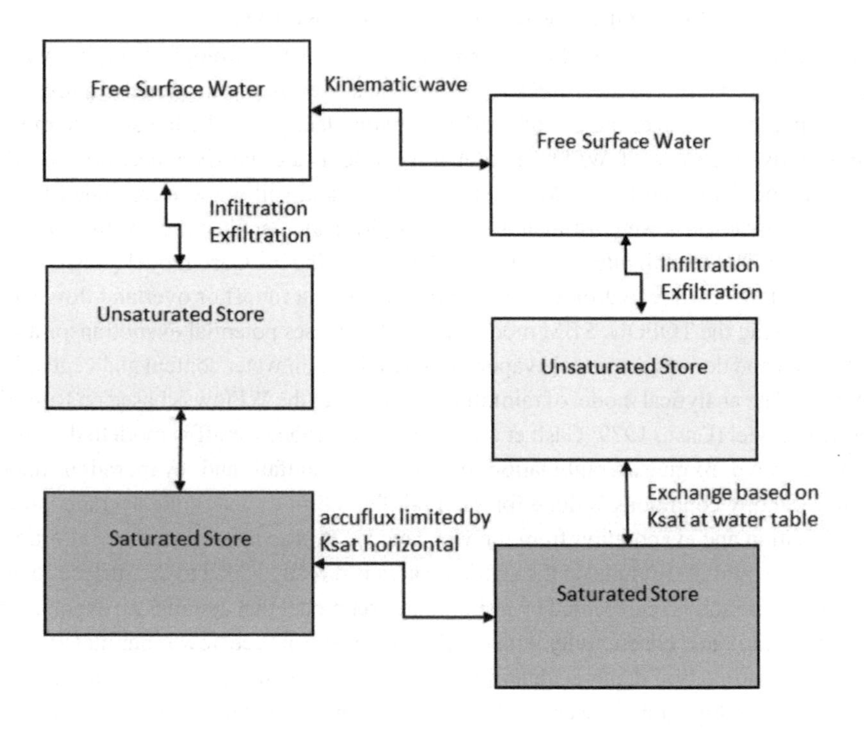


*Figure 4.1: Schematization of the soil within the WFlow_sbm model*
*Source: http://WFlow.readthedocs.io/en/latest/WFlow_sbm.html/the-soil-model.*

### 4.4.1 Model calibration and validation

As with all hydrological models, calibration of the Dinder and Rahad hydrological model is needed for optimal performance. Since the hydrological data available for calibration start from 2001, the nearest land use (land use data from 1998) was used in the calibration. The calibration procedure performed in two steps based on, first, initial values of all parameters were estimated based on the land use and the soil types. Second, by adjusting the model parameters and evaluate the results.

The performance of the model was assessed using measures of goodness of fit between the modelled and observed flow using the coefficient of determination ($R^2$) and the Nash–Sutcliffe efficiency (NSE), defined by Nash and Sutcliffe (1970). The observed and the simulated flow of the Dinder and Rahad correlated well, except for few underpredictions and overpredictions of peak flows, which can be explained in terms of inherent uncertainty in the model and the data. However, measures of performances for both calibration and verification runs fell within the acceptable ranges.

## 4.5 INDICATORS OF HYDROLOGIC ALTERATIONS (IHA)

The IHA approach was introduced by Richter et al. (1996). The approach used to assess river ecosystem management objectives which defined based on a statistical representation of the most ecologically relevant hydrologic indicators. These indicators describe the essential characteristics of a river flow that have ecological implications. The IHA method computes 32 hydrologic parameters for each year. For analyzing the alteration between two periods, the IHA described in Richter et al. (1996) was applied using the IHA software developed by The Nature Conservancy (2009).

The general approach is to define hydrologic parameters that characterized the intra-annual variation in the water system condition and then to use the analysis of variations in these parameters as a base for comparing hydrologic alterations of the system before and after the system has been altered by various human activities.

The IHA method has four steps: (a) define the time series of the hydrologic variable (e.g. streamflow) for the two periods to be compared, (b) calculate values for hydrologic parameters, (c) compute intra-annual statistics and (d) calculate values of the IHA by comparing the intra-annual variation before and after the system has been altered and present the results as a percentage of deviation. For assessing hydrologic alteration in the Dinder and Rahad rivers, the flow variations for both rivers have been characterized based on the variations in the streamflow characteristics between three periods: 1972–1986, 1986–1998 and 1998–2011. Temporal variability of streamflow series was analyzed at Al

Gewisi station on the Dinder river and at Al Hawata station on the Rahad river. A detailed description of IHA can be found in Richter et al. (1996) and Poff et al. (1997).

## 4.6 Results and Discussion

### 4.6.1 LULC classification and change detection

The overall LULC classification accuracy levels for the four images ranged from 82% to 87%, with Kappa indices of agreement ranging from 77% to 83% (Table 4.3). The accuracy assessment is based on comparing reference data (class types at specific locations from ground information) to image classification results at the same locations. The overall accuracy of classification is the average value from all classes. The user's accuracy corresponds to errors of inclusion (commission errors), which represents the probability of a pixel classified into a given class actually representing that class on the ground (i.e. from the perspective of the user of the classified map). The producer's accuracy corresponds to errors of exclusion (omission errors), which represents how well reference pixels of the ground cover type are classified (i.e. from the perspective of the maker of the classified map). The commission errors occur when an area is included in an incorrect category, while the omission errors occur when an area is excluded from the category to which it belongs. Every error on the map is an omission from the correct class and a commission to an incorrect class (Congalton and Green, 2008). The cross-validation for the land use data from year 2011 was made using the reference data (120 points) collected with a GPS instrument during the field survey (2011–2013). In addition, visual interpretation and historical information obtained from the local people about the land use types in the study area were also used as cross-check validations for old maps. Shrub lands show lower user's and producer's accuracies compared to the other LULC classes. This is mainly due to the mis-classification of some shrub land into woodland, grassland and cropland. This accuracy is satisfactory for the study area considering the multi-temporal analysis of Landsat data and the visual interpretation adapted to image classification.

*Table 4.3: Accuracy assessment (%) of LULC maps.*

| LULC Classes | 1972 | | 1986 | | 1998 | | 2011 | |
|---|---|---|---|---|---|---|---|---|
| | Producer's | User's | Producer's | User's | Producer's | User's | Producer's | User's |
| Woodland | 88 | 89 | 89 | 90 | 89 | 90 | 91 | 93 |
| Cropland | 78 | 70 | 80 | 74 | 80 | 80 | 83 | 82 |
| Shrubland | 71 | 71 | 73 | 75 | 77 | 75 | 80 | 75 |
| Grassland | 80 | 88 | 83 | 88 | 85 | 88 | 86 | 89 |
| Bare Land | 82 | 76 | 82 | 78 | 82 | 78 | 82 | 85 |
| Water | 86 | 86 | 88 | 86 | 91 | 86 | 94 | 86 |
| Overall | 82 | | 84 | | 85 | | 87 | |
| Kappa | 77 | | 79 | | 81 | | 83 | |

Landsat image classification results for the years 1972, 1986, 1998 and 2011 are shown in Figure 4.2. The large extent of the catchment (77,504 km$^2$), and the small-scale of the maps (i.e. 1:4 500 000), may not allow distinction of different LULC change patterns by eye. Figure 4.3, which shows a closer view of the area, is an example to show multi-temporal changes in the LULC patterns. The zoomed-in areas in the red boxes, shown on a large scale, provide more details of the LULC patterns. This area is located downstream of the Rahad irrigation scheme of Sudan, established in 1981. The waterlogging and woodland areas that occurred in 1998 and 2011 resulted from the drainage water of the project accumulating over the years (a clear example of LULC multi-temporal change over the Rahad basin). The lower maps show the Google Earth images of the large-scale area. Although the dates of these Google Earth images do not exactly match the ones of the satellite images, they show the part of the dried period in the study area and hence the complexity of the LULC patterns.

According to the produced LULC maps, it was found that woodland, shrub land and grassland were the dominant types of LULC classes for the years 1972, while for the year 1986 they were shrub land, grassland and cropland. The LULC map of 1998 illustrates that the predominant types of LULC classes were cropland and woodland, while they were cropland and shrubs in 2011.

LULC changes in the D&R are assessed by image comparison. In general, the results showed that the dominant process is the large decrease in woodland and increase in cropland. This result was in agreement with that of Rientjes et al. (2011) and Gumindoga et al. (2014), who studied the changes in land cover, rainfall and streamflow in the neighboring catchment of the upper Gilgel Abay in Ethiopia.

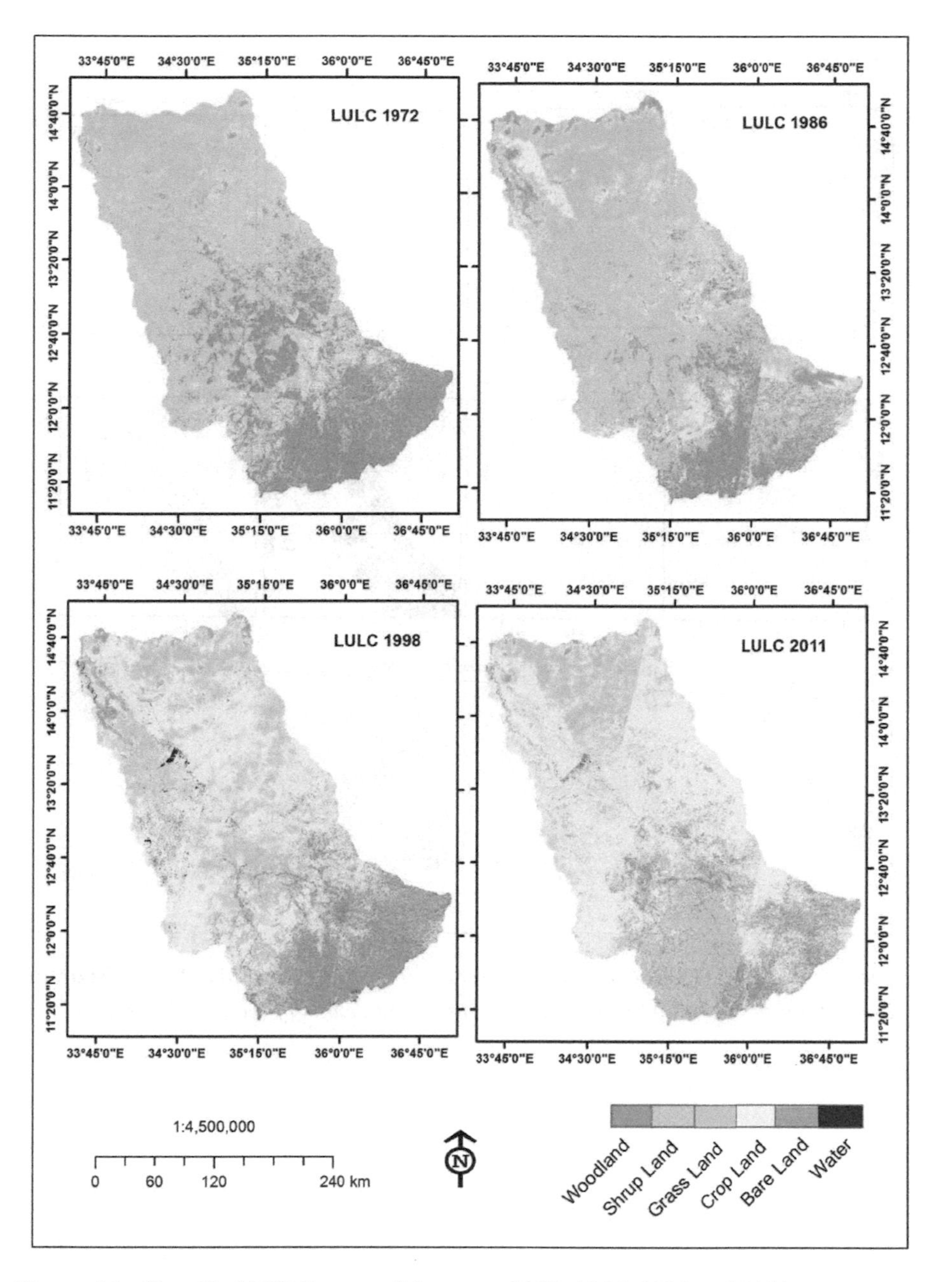


*Figure 4.2: Classified LULC maps of the years 1972, 1986, 1998 and 2011.*

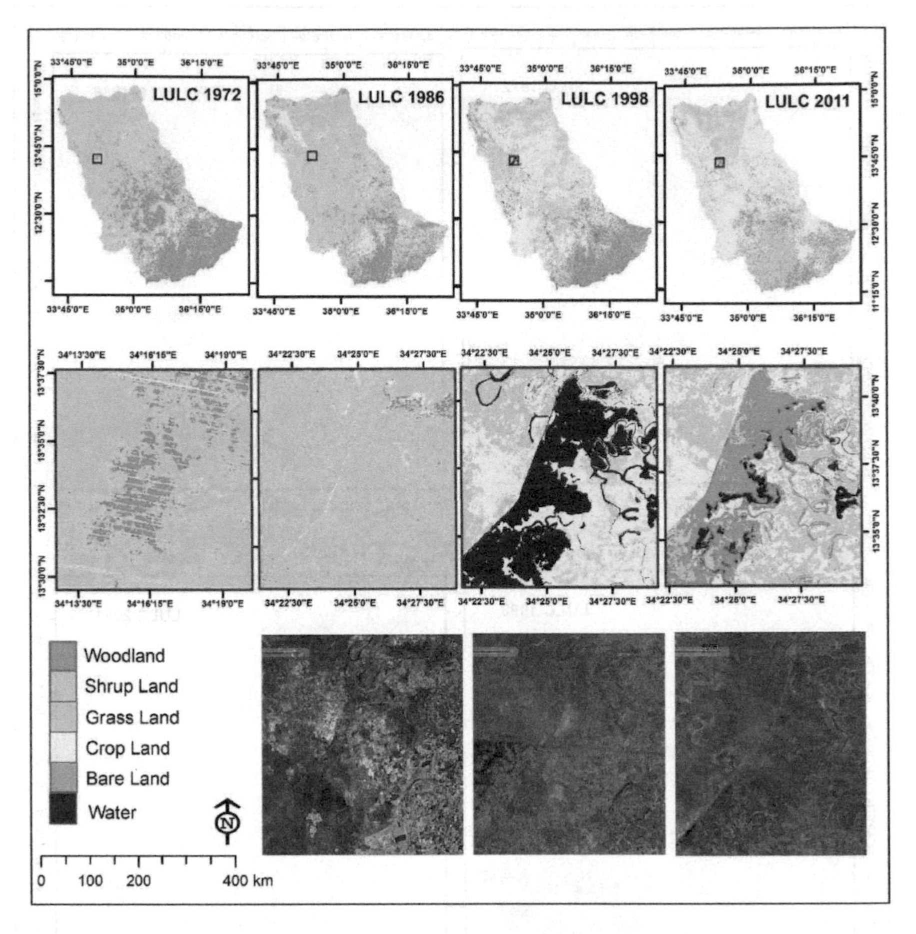


*Figure 4.3: Classified LULC maps of the years 1972, 1986, 1998 and 2011. The areas in the red boxes showed in large-scale provide more details of the LULC changes patterns.*

Table 4.4 shows the percentages of LULC changes classes in Dinder and Rahad basins that occurred in the periods 1972–1986, 1986–1998 and 1998–2011. The decrease in the woodland area in 1986 is mainly attributed to the deforestation during the drought time in 1984 and 1985. As a result, the cropland was increased due to the development of new agricultural areas in both irrigated (i.e. Rahad irrigation scheme) and rain-fed sectors. The rapid expansion in the mechanized rain-fed agriculture led to a large increase in cropland during 1998 and 2011. These findings are in agreement with what has been reported by Marcotullio and Onishi (2008) and Biro et al. (2013) from their similar studies conducted in the Ethiopian highlands and the Gedarif region in eastern Sudan.

*Table 4.4: Land cover (%) in Dinder and Rahad basins that occurred in the period 1972 to 1986, 1986 to 1998, and 1998 to 2011.*

| Land cover type (%) | Dinder | | | | Rahad | | | |
|---|---|---|---|---|---|---|---|---|
| | 1972 | 1986 | 1998 | 2011 | 1972 | 1986 | 1998 | 2011 |
| Bare area | 5 | 1 | 0 | 2 | 6 | 5 | 0 | 3 |
| Woodland | 42 | 23 | 27 | 14 | 35 | 14 | 21 | 14 |
| Shrubland | 23 | 43 | 21 | 36 | 30 | 32 | 13 | 15 |
| Grassland | 16 | 18 | 5 | 1 | 11 | 22 | 9 | 1 |
| Cropland | 14 | 15 | 45 | 47 | 18 | 26 | 55 | 68 |

### 4.6.2 Calibration and validation of the hydrological model results

To assess the reliability of the SBRE products, validation is carried out with the use of ground measurements at four gauges in which observed data are available. Two gauges (Gonder and Bahir Dar) are located nearby the upstream part of the catchments in the Ethiopian plateau, while the other two (Gedarif and Al Hawata) are located at the most downstream part of the catchment in the Sudanese lowland. The validation is performed at annual time step. The results show that the difference of RFE against ground measurements has no consistent patterns. TRMM and CHIRPS have shown no consistent patterns at the lowland (Gedarif and Al Hawata), but both products are consistent and overestimate rainfall at the Ethiopian highland (Gonder and Bahir Dar) in all years except 2007 (Figure 4.4). Since both the Dinder and the Rahad derive their main flow from the

Ethiopian highlands, products with consistent patterns in the highlands will be more suitable for running hydrologic models in this catchment. From these findings, one can conclude that the CHIRPS v2.0 and TRMM 3B42 v7 are more suitable than RFE 2.0 for running hydrologic model. Comparing CHIRPS v2.0 and TRMM 3B42 v7, it is clear that CHIRPS v2.0 has less overestimation of rainfall. Thus, CHIRPS v2.0 is the best product to be used as a forcing data for hydrologic model in the Dinder and Rahad basins.

The NSE and $R^2$ ranged from 0.4 to 0.80 and 0.50 to 0.80, respectively, for both the daily calibration and validation for the three precipitation products at Al Gewisi station on the Dinder river and Al Hawata station on the Rahad river (Figure 4.5 and 4.6). At Al Gewisi station, the large underestimation in the first validation period for CHIRPS can be attributed to the underestimation of rainfall by CHIRPS in 2007 at both Gonder and Bahir Dar (Figure 4.4), while at the same time CHIRPS overestimates rainfall in all years from 2001 to 2006. Therefore, calibration of the hydrologic model (during the period 2002–2005) resulted in underestimation of river flow in 2007. On the other hand, at Al Hawata station, the difference between observed and model flow in the first period of validation (i.e. 2008) is likely due to an error either in the input data or the observed flow values or a combination of both.

In general, the calibration results indicate that CHIRPS 2.0 is the best product over rugged terrains with complex rainfall patterns, such as those in the D&R basins. This result is in agreement with Hessels (2015), who compared and validated 10 open-access and spatially distributed satellite rainfall products over the Nile Basin and found that CHIRPS is the best product to be used in the Nile Basin. The modelling results show that the approach is reasonably good and therefore can be used to predict runoff at a sub-basin level. Then the model was used to simulate the impact of LULC changes on streamflow by running the model using land cover from different periods of time (1972, 1986, 1998 and 2011) and keeping precipitation (CHIRPS), evapotranspiration and other model parameters without change.

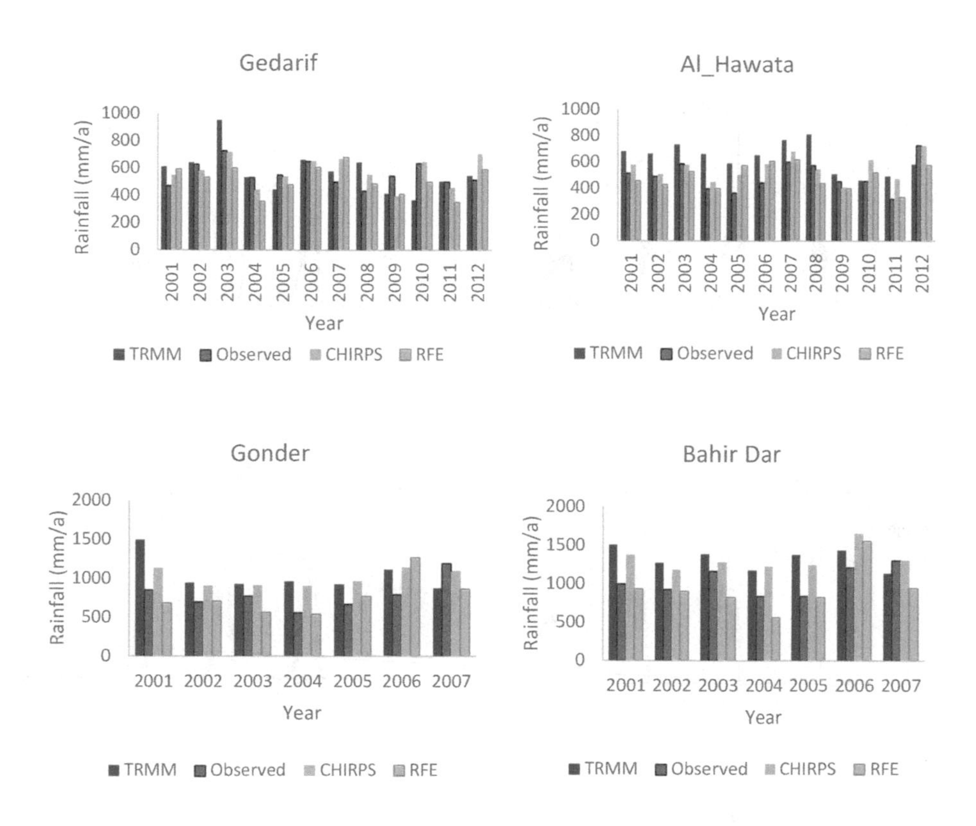


*Figure 4.4: Comparison of SBRE products with ground measurements at four locations.*

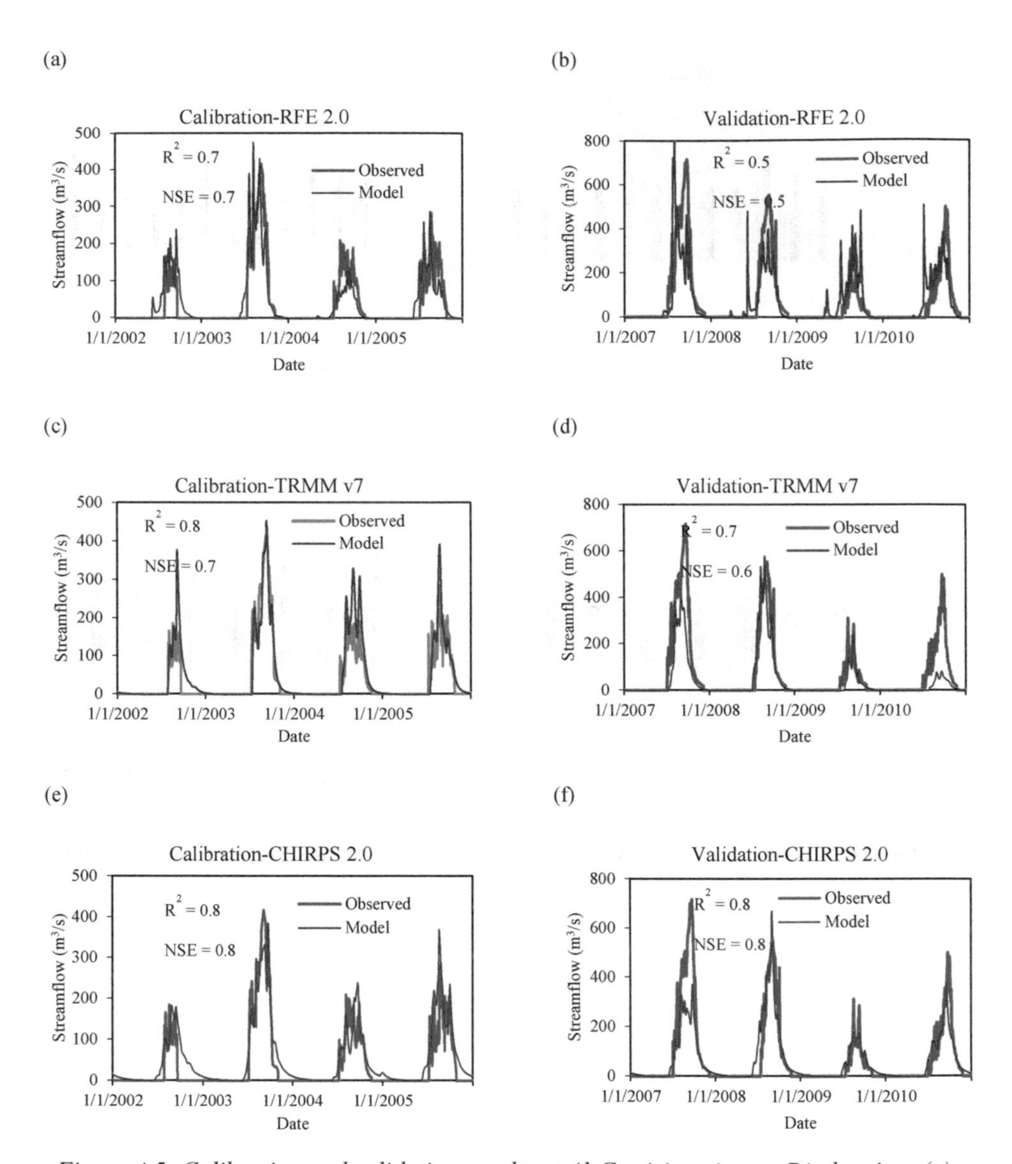


*Figure 4.5: Calibration and validation results at Al-Gewisi station on Dinder river (a) and (b) for RFE, (c) and (d) for TRMM and (e) and (f) for CHIRPS.*

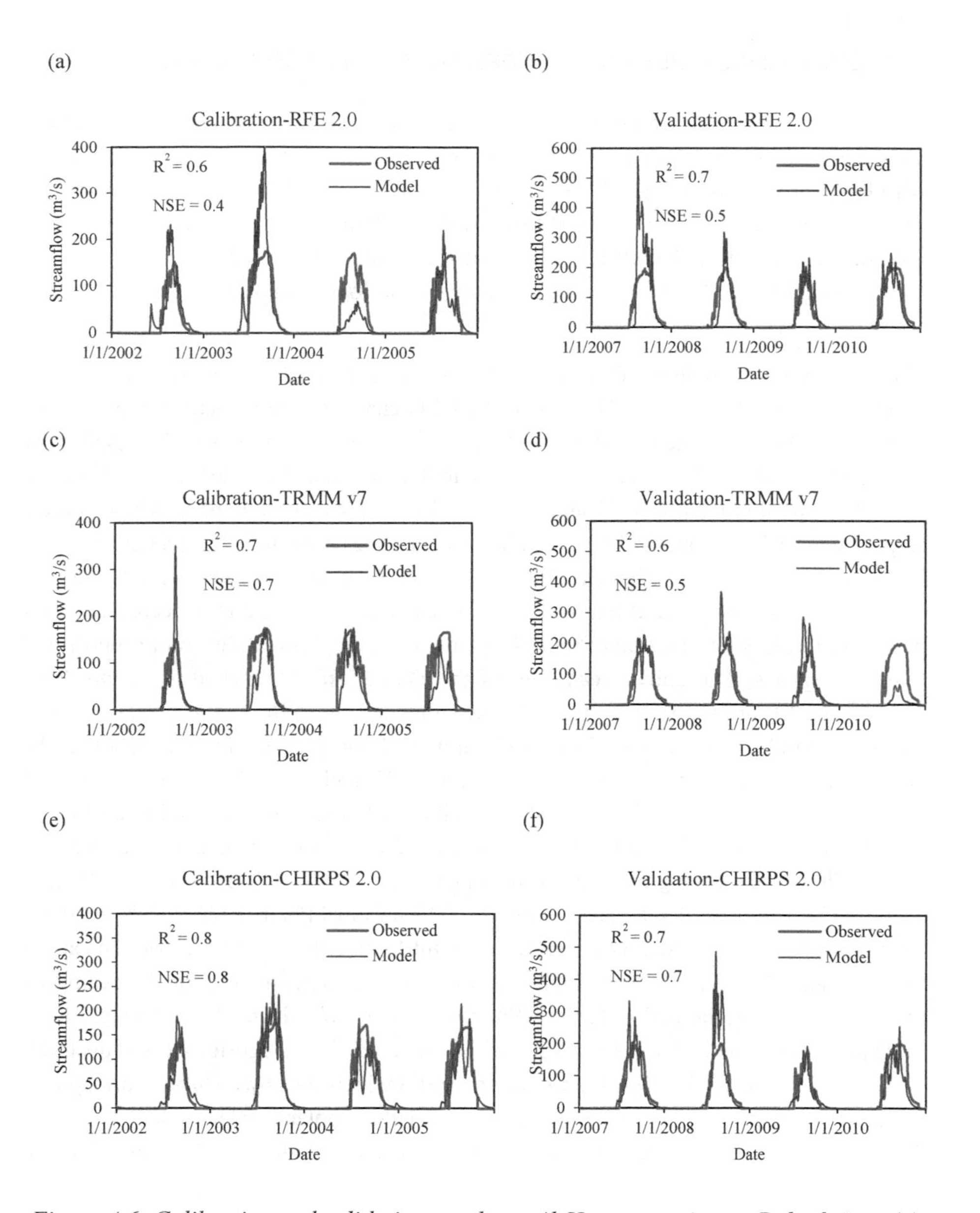


*Figure 4.6: Calibration and validation results at Al-Hawata station on Rahad river (a) and (b) for RFE, (c) and (d) for TRMM and (e) and (f) for CHIRPS.*

## 4.7 STREAMFLOW RESPONSE UNDER LAND COVER CONVERSIONS

After the calibration and validation of the WFlow, the model has been run using different land use data with fixed model parameters: first, with land use data from 1972; second, with land use data from 1986; third, with land use data from 1998; and fourth, with land use data from 2011. Then the output flows from the four land uses were compared. We note that the rainfall (CHIRPS) and PET for the period 2001–2012 were used with the 1972, 1986, 1998 and 2011 land uses to identify hydrological impacts of changes in land cover explicitly.

The WFlow result indicates that streamflow is affected by LULC changes in both the Dinder and the Rahad rivers. The effect of LULC changes is much larger in the Rahad than in the Dinder. In the Rahad basin, the simulated streamflow showed low peak flow with land use data from 1972 and high flow with land use data from 2011. Woodland and shrub land are dominant in 1972 and occupied 35 and 30% of the upper catchment area, respectively, while cropland is the dominant land cover type in 2011, occupying 68%. Woodland and shrub land have high porosity and delayed the release of water to the catchment outlet. Woodland removal implies less infiltration due to a decrease in soil permeability, less interception of rainfall by the tree canopies and thus more runoff and high flow peaks. The daily streamflow of the Dinder and the Rahad as results from different LULC are shown in Figure 4.7. The simulated streamflow of the Rahad river as a result of land covers of 1972, 1986, 1998 and 2011 was presented in Figure 4.8b. The annual streamflow increased by 75% between 1972 and 1986, but is followed by a decrease of 45% between 1986 and 1998. The increase in streamflow could be a result of a decrease in woodland by 60% from 35% in 1972 to 14% in 1986, associated with an increase in cropland and grassland. Cropland has increased by 44% from 18% in 1972 to 26% in 1986 and grassland has increased by 100% from 11% in 1972 to 22% in 1986. This increase in grassland thus decreases water infiltration due to soil compaction caused by grazing, which causes both higher runoff and an increase in annual streamflow magnitude. During the period 1986–1998, cropland and woodland showed a significant increase by 113% and 53%, respectively, while the remaining categories showed declines. During the period 1998–2011, the annual streamflow increased by 65% and corresponds with results on increases in the percentage of bare land, cropland and shrubland by 754%, 23% and 15%, respectively, while a decrease in woodland and grassland by 37% and 94%, respectively.

Similar to the Rahad, the simulated streamflow of the Dinder river showed low peak flow with land use data from 1972 and relatively high flow with land use data from 2011. Woodland is dominant in 1972 and occupied 42% of the total catchment area, while cropland is the dominant land cover type in 2011, occupying 47%. Figure 4.8a shows the simulated annual streamflow of the Dinder river as a result of land cover data of 1972,

1986, 1998 and 2011. Annual stream-flow increased by 20% between 1972 and 1986 but is followed by a decrease of 9% between 1986 and 1998. This could be a result of a decrease in woodland by 43% from 42% in 1972 to 23% in 1986 associated with an increase in shrub land, grassland and cropland by 83%, 10% and 6%, respectively. During the period 1986–1998, cropland and woodland increased by 192% and 16%, respectively, while the remaining categories showed declines. Over the period 1998–2011, the annual streamflow increased by 52% and corresponds with findings on increases in the percentage of bare land, cropland and shrub land by 360%, 4% and 71%, respectively, while a decrease in woodland and grassland by 50%, and 76%, respectively. The decrease in percentage change of bare area over the period 1986–1998, along with the increase in woodland in both the Dinder and the Rahad basins, indicates that the environment was recovering from the severe drought of 1984–1985.

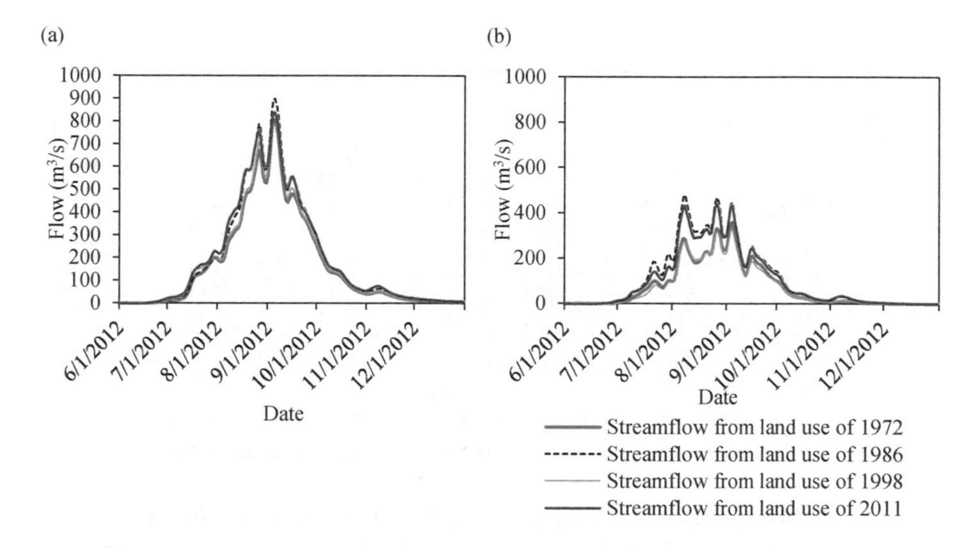


*Figure 4.7: Daily streamflow results from the WFlow model at (a) Al-Gewisi station on the Dinder river and (b) Al-Hawata station on the Rahad river based on land use from 1972, 1986, 1998 and 2011 for the year 2012 as an example.*

(a)

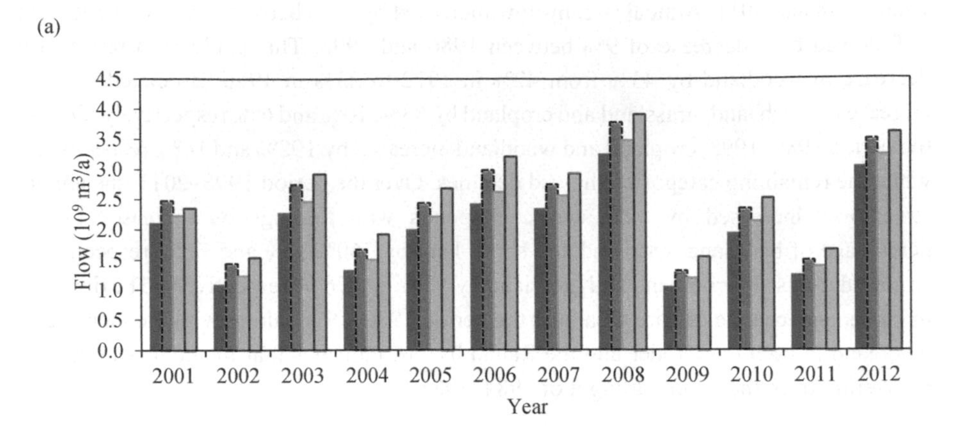


(b)

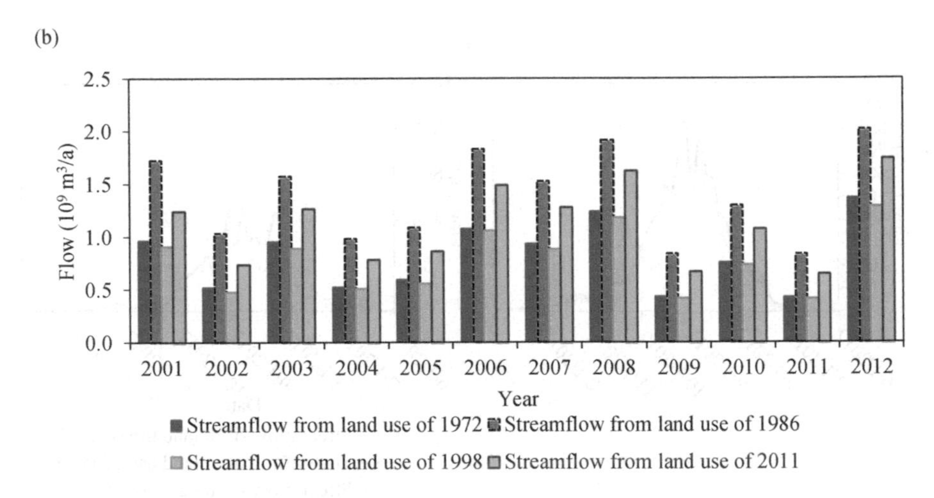


*Figure 4.8: Annual streamflow results from the WFlow model at (a) Al-Gewisi station on the Dinder river and (b) Al-Hawata station on the Rahad river based on land use from 1972, 1986, 1998 and 2011.*

In addition to the streamflow response to LULC changes, evapotranspiration (ET) is another important component of the water balance that constitutes a major determinant of the amounts of water draining from different land cover types within the catchment. The ET result shows high rates of actual evapotranspiration (AET) when running the model with land cover data from the years 1972 and 1998 at both the sub-catchments and over the entire catchment (Tables 4.5 and 4.6).

*Table 4.5: Annual average AET as a response to LULC changes at the sub-catchments level for the Dinder catchment (1972-1986).*

| | AET from land cover of 1972 (mm/a) | | | | AET from land cover of 1986 (mm/a) | | | |
|---|---|---|---|---|---|---|---|---|
| Year | Al-Gewisi | Musa | Gelagu | Upper Dinder | Al-Gewisi | Musa | Gelagu | Upper Dinder |
| 2001 | 558 | 583 | 626 | 464 | 426 | 424 | 396 | 288 |
| 2002 | 443 | 456 | 535 | 510 | 322 | 317 | 306 | 312 |
| 2003 | 564 | 639 | 642 | 486 | 425 | 469 | 405 | 312 |
| 2004 | 455 | 502 | 573 | 500 | 326 | 354 | 340 | 311 |
| 2005 | 504 | 547 | 575 | 505 | 376 | 396 | 358 | 323 |
| 2006 | 527 | 576 | 632 | 545 | 396 | 414 | 406 | 359 |
| 2007 | 598 | 602 | 618 | 564 | 468 | 444 | 400 | 382 |
| 2008 | 593 | 689 | 703 | 576 | 459 | 513 | 471 | 392 |
| 2009 | 421 | 482 | 519 | 516 | 310 | 343 | 302 | 323 |
| 2010 | 536 | 566 | 606 | 520 | 412 | 415 | 383 | 331 |
| 2011 | 470 | 467 | 554 | 530 | 350 | 327 | 329 | 332 |
| 2012 | 636 | 679 | 684 | 542 | 500 | 504 | 450 | 353 |

*Table 4.6: Water balance of the Dinder and Rahad catchments applying different LULC.*

| Dinder catchment | | Land cover of 1972 | | Land cover of 1986 | | Land cover of 1998 | | Land cover of 2011 | |
|---|---|---|---|---|---|---|---|---|---|
| Year | Rainfall (mm/a) | AET (mm/a) | Streamflow (mm/a) | AET (mm/a) | Streamflow (mm/a) | AET (mm/a) | Streamflow (mm/a) | AET (mm/a) | Streamflow (mm/a) |
| 2001 | 816 | 558 | 258 | 383 | 433 | 432 | 384 | 496 | 320 |
| 2002 | 663 | 486 | 177 | 314 | 349 | 364 | 299 | 430 | 233 |
| 2003 | 847 | 583 | 264 | 403 | 444 | 449 | 397 | 519 | 327 |
| 2004 | 703 | 507 | 195 | 333 | 370 | 374 | 329 | 451 | 252 |
| 2005 | 768 | 532 | 236 | 363 | 405 | 414 | 354 | 479 | 289 |
| 2006 | 835 | 570 | 265 | 394 | 441 | 441 | 395 | 513 | 322 |
| 2007 | 876 | 595 | 280 | 424 | 452 | 476 | 400 | 540 | 336 |
| 2008 | 929 | 640 | 289 | 459 | 470 | 509 | 420 | 582 | 347 |
| 2009 | 659 | 484 | 175 | 319 | 340 | 363 | 297 | 435 | 225 |
| 2010 | 817 | 557 | 260 | 385 | 432 | 432 | 386 | 505 | 312 |
| 2011 | 710 | 505 | 205 | 334 | 376 | 377 | 333 | 454 | 256 |
| 2012 | 972 | 635 | 337 | 452 | 520 | 498 | 474 | 579 | 393 |
| Rahad catchment | | Land cover of 1972 | | Land cover of 1986 | | Land cover of 1998 | | Land cover of 2011 | |
| Year | Rainfall (mm/a) | AET (mm/a) | Streamflow (mm/a) | AET (mm/a) | Streamflow (mm/a) | AET (mm/a) | Streamflow (mm/a) | AET (mm/a) | Streamflow (mm/a) |
| 2001 | 724 | 409 | 315 | 290 | 434 | 398 | 326 | 309 | 416 |
| 2002 | 641 | 398 | 243 | 271 | 370 | 383 | 258 | 291 | 350 |
| 2003 | 755 | 450 | 305 | 323 | 432 | 434 | 322 | 342 | 413 |
| 2004 | 609 | 360 | 249 | 231 | 378 | 338 | 270 | 244 | 364 |
| 2005 | 656 | 399 | 258 | 267 | 389 | 378 | 278 | 285 | 372 |

| | | | | | | | | | |
|---|---|---|---|---|---|---|---|---|---|
| 2006 | 782 | 450 | 332 | 324 | 457 | 431 | 351 | 336 | 446 |
| 2007 | 774 | 473 | 301 | 344 | 430 | 456 | 319 | 363 | 411 |
| 2008 | 754 | 438 | 315 | 313 | 441 | 415 | 338 | 322 | 431 |
| 2009 | 581 | 352 | 229 | 220 | 361 | 333 | 248 | 238 | 343 |
| 2010 | 744 | 449 | 295 | 319 | 425 | 431 | 313 | 335 | 409 |
| 2011 | 610 | 369 | 241 | 235 | 375 | 348 | 262 | 252 | 358 |
| 2012 | 873 | 507 | 366 | 381 | 492 | 485 | 388 | 390 | 483 |

This can be attributed to the large percentage coverage of woodland in 1972 and 1998 compared to land cover data from 1986 and 2011 (please refer to Table 4.4). The lowest AET is observed when running the model with land cover data from 1986. This is likely due to the severe drought during the mid-1980s that limits the water availability and decreases the green coverage. Table 4.5 presents the change in the annual average AET at sub-catchment level as a response to LULC changes for the Dinder catchment. Table 4.6 shows the changes in water balance for the entire Dinder and Rahad catchments when running the hydrologic model with different LULC and fixed rainfall data for the periods 2001–2012.

Since both the Dinder and the Rahad rivers are seasonal, their flows mainly depend on rainfall patterns and magnitudes. In addition to the effect of LULC changes on the streamflow, Figure 4.9 shows that the annual variability of rainfall is another factor affecting the annual patterns of the streamflow.

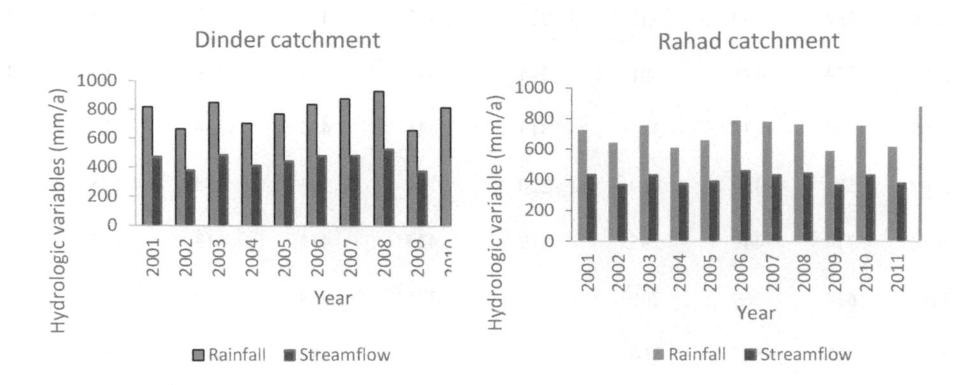


*Figure 4.9: Annual average rainfall and streamflow patterns and magnitudes for the years (2001-2012).*

## 4.8 STREAMFLOW ANALYSIS WITH IHA

Since both Dinder and Rahad are seasonal rivers (July– November) and its floodplains, including the mayas, are mainly depending on floods, the streamflow analysis is focused on the flows during the months of high flows and the indicators describing the hydrological high extremes. The investigated streamflow variables are a subset of the 32 indicators proposed by Richter et al. (1996) under the range of variability approach (RVA) that characterizes the natural flow regime of a river into five categories of magnitude, timing, duration, frequency and rate of change. In this section, the modelled streamflow as a result of LULC data from 1972, 1986, 1998 and 2011 has been analyzed.

### 4.8.1 Magnitude of monthly flow

The general pattern of median monthly flow of the Rahad river (Figure 4.10a) at Al Hawata station during 1972–1986 is that the median flow increased in all months of flow (July– November) with an average of 83% per month. In contrast, the median monthly flow decreased in all months during the period 1986–1998 with an average of 45% per month. Similar to the period 1972–1986, the median monthly flow during 1998–2011 increased by an average of 65% per month. In comparison to Rahad, the Dinder median monthly flow (Figure 4.10b) at Al Gewisi station during 1972–1986 increased in all months of flow by an average of 21% per month. In contrast, the median monthly flow decreased in all months during the period 1986–1998 with an average of 6% per month. Likewise, to the period from 1972–1986, the median monthly flow during 1998–2011 increased by an average of 17% per month. Alterations of the monthly flow magnitude, particularly during the months of high flows (August–October) is likely affecting habitat availability on floodplains, which may lead to decreases and/or disappearance of native flora and increases in non-native flora that might not be suitable for the herbivorous wildlife that dwells in the DNP.

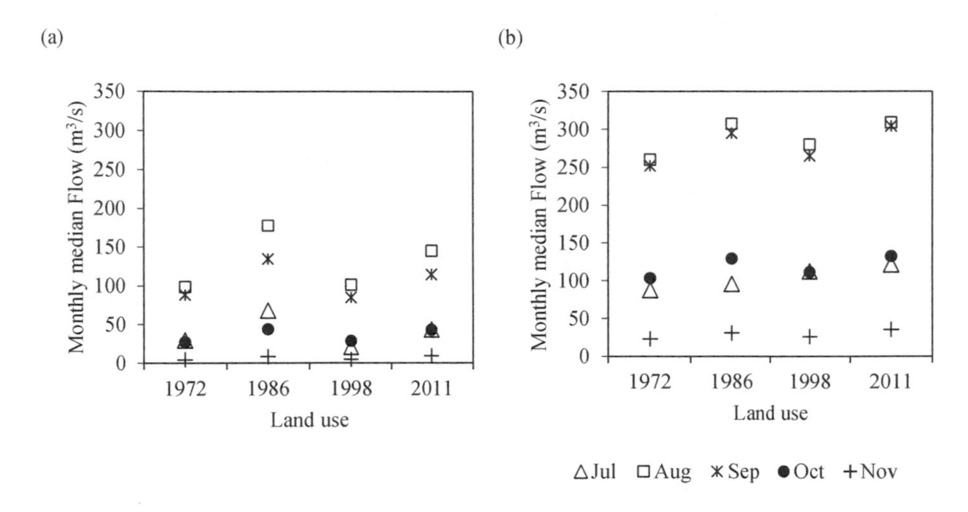


*Figure 4.10: The monthly median flow (a) for Rahad river and (b) for Dinder river.*

## 4.8.2 Magnitude of river extreme floods

Extreme floods are important in re-forming both the biological and physical structure of a river and its associated floodplain. Extreme floods are also important in the formation of key habitats such as oxbow lakes and floodplain wetlands. The pattern of the extreme flow is vital for the filling of maya wetlands of the DNP. Therefore, annual flow maxima of 1, 7, 30 and 90-day intervals have been investigated. The median maxima are presented in Figure 4.11. In general, all results have shown that the maxima are significantly affected by LULC changes. In Rahad, median flow maxima for 1, 7, 30 and 90-day intervals from the land use data from 1986 are 51%, 56%, 67% and 68%, respectively, higher than the maxima from the land use data from 1972. Likewise, median flow maxima for 1, 7, 30 and 90-day intervals from the land use data from 2011 are 32%, 33%, 36% and 39%, respectively, higher than the maxima from the land use data from 1998. In contrast, median flow maxima for 1, 7, 30 and 90-day intervals from the land use data from 1998 are 39%, 39%, 42%, and 42%, respectively lower than the maxima from the land use data from 1986.

In the Dinder river the effect of LULC changes on streamflow is not big as in Rahad river. This is likely due to the large expansion in cropland in the Rahad catchment to 68% of the total area compared to 47% in the Dinder catchment. The median flow maxima for 1, 7, 30 and 90-day intervals from the land use data from 1986 are 19%, 19%, 18% and 18%, respectively, higher than the maxima from the land use data from 1972. Likewise, the median flow maxima for 1, 7, 30 and 90-day intervals from the land use data from 2011 are 14%, 13%, 14% and 19% respectively, higher than the maxima from the land use data from 1998. In contrast, the median flow maxima for 1, 7, 30 and 90-day intervals from the land use data from 1998 are 11%, 11%, 10% and 10%, respectively, lower than the maxima from the land use data from 1986. Peak flows are the critical aspects of the lateral connectivity between the Rahad and the Dinder rivers and its floodplains. Reduction of the magnitude of these high-flow peaks during dry years (less than average) may reduce the ecological function of the maya wetlands areas as breeding, nursery and feeding habitat for wildlife.

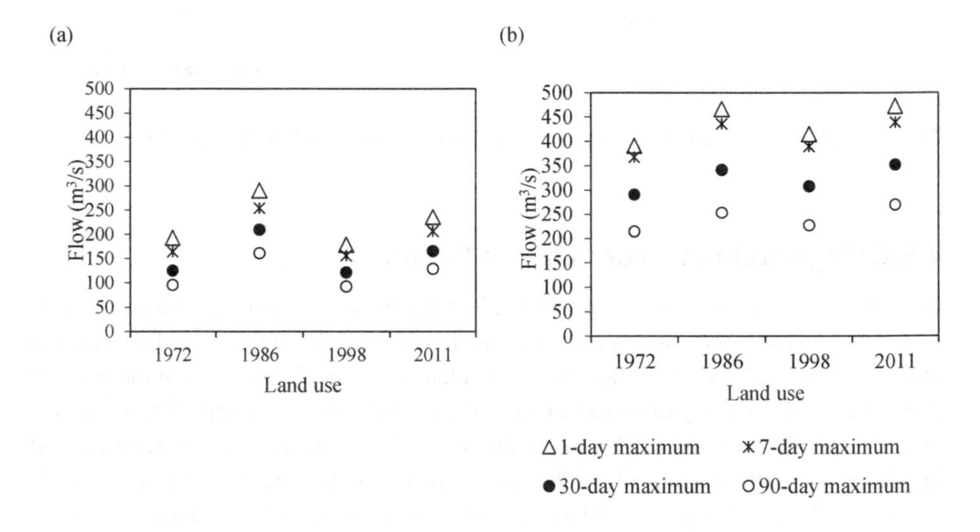


*Figure 4.11: Median flow maxima for 1, 7, 30 and 90-day intervals from the land use of 1972, 1986, 1998 and 2011 for (a) Rahad river and (b) Dinder river.*

### 4.8.3 Timing of annual extreme floods

Synchronization of annual flooding with a variety of riverine and floodplain species life-cycle requirements is likely to be of high importance given the adaptation of species to their habitat. In the Rahad river, dates of the annual maxima as results from the land use data from 1972, 1986, 1998 and 2011 occurred within the same three weeks (15$^{th}$ August–2$^{nd}$ September, Julian date (JD) 227–245). The annual maxima from the land use data from 1986 is 18 days earlier than the annual maxima from land use data from 1972. This could be attributed to land cover degradation and deforestation due to the devastating drought of 1984–1985 resulting in acceleration of the runoff response. In Dinder river, dates of the annual maxima are not affected by LULC changes and occurred within the same two days (11– 12 September, JD 254–255).

### 4.8.4 Rate of change in flow

The rate of change in flow can affect persistence and lifetime for both aquatic and riparian species (Poff et al. 1997), particularly in arid areas where streamflow usually changes rapidly in a very short time. Figure 4.12 shows the rate of flow rises and flow falls for both Rahad and Dinder. The median rate of flow rises (positive differences between consecutive daily values) in Rahad river has increased by 74% from 2.73 ($m^3$/s) /day in 1972 to 4.73 ($m^3$/s) /day in 1986. In 1998 the median rate of flow rises decreased by 50%, while increasing by 37% in 2011. Similarly, the median rate of flow falls (negative differences between consecutive daily values) has increased by 88% from 0.12 ($m^3$/s) /day in 1972 to 0.23 ($m^3$/s) /day in 1986. In 1998 the median rate of flow falls decreased by 37%, while increasing by 22% in 2011. Likewise, the median rate of flow rises and flow falls in the Dinder river follows the same pattern of the Rahad flow, but no significant changes were observed. This result shows that the fluctuation in rate of change in streamflow is strongly linked to LULC changes, especially when analyzing the streamflow as a result of land use after a period of drought (e.g. land use data from 1986).

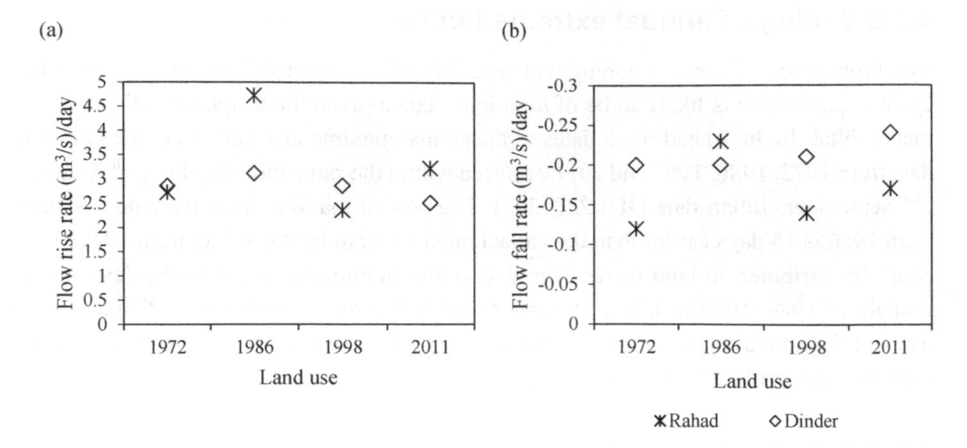


*Figure 4.12: The rate of flow rises (a) and falls (b) as a response to land use of 1972, 1986, 1998 and 2011 for both Rahad and Dinder rivers (negative sign in the vertical axis indicates downward direction of flow).*

## 4.9 CONCLUSIONS

For assessing the changes in land cover, four remote sensing images were used for the years 1972, 1986, 1998 and 2011. The accuracy assessment with supervised land cover classification shows that the classification results are reliable. The land cover changes in the D&R are assessed by image comparison and the results showed that the dominant process is the relatively large decrease in woodland and the large increase in cropland. The results of LULC changes detection between 1972 and 2011 indicate a significant decrease in woodland and an increase in cropland. Woodland decreased from 42% to 14% and from 35% to 14% for Dinder and Rahad, respectively. Cropland increased from 14% to 47% and from 18% to 68% in Dinder and Rahad, respectively. The rate of deforestation is high during the period 1972–1986 and is probably due to the severe drought during 1984–1985 and expansion of agricultural activities as well as increased demand for wood for fuel, construction and other human needs due to the increase in population. On the other hand, the increase in woodland during the period between 1986 and 1998 is probably due to reforestation activities in the basin. Nevertheless, the magnitude of deforestation is still much larger than the reforestation. The cropland expansion over the period 1986– 1998 is larger than the expansion over the period 1998–2011, suggesting that most of the areas that are suitable for cultivation have most likely been occupied, or the land tenure regulations have controlled the expansion of cultivation by local communities.

The results of the hydrological model indicate that stream-flow is affected by LULC changes in both the Dinder and the Rahad rivers. The effect of LULC changes on streamflow is significant during 1986 and 2011, particularly in the Rahad river. This could be attributed to the severe drought during 1984–1985 and the large expansion in cropland in the Rahad catchment to 68% of the total area.

The IHA-based analysis indicated that the flows of the Dinder and the Rahad rivers were associated with significant upward and downward alterations in magnitude, timing and rate of change of river flows, as a result of LULC changes. These alterations in the streamflow characteristics are likely to have significant effects on a range of species that depend on the seasonal patterns of flow. Therefore, alterations in the magnitude of the annual floods that decrease the water flowing to the mayas may reduce the production of native river floodplain fauna and flora and the migration of animals that arc connected to mayas inundation.

The effects of hydrological and morphological changes on mayas inundation are discussed in the next chapter.

# 5

# MODELLING THE INUNDATION AND MORPHOLOGY OF THE SEASONALLY FLOODED MAYAS WETLANDS IN THE DINDER NATIONAL PARK

[5] This chapter is based on: Hassaballah, K., Y. Mohamed, A. Omer & S. Uhlenbrook, 2020. Modelling the Inundation and Morphology of the Seasonally Flooded Mayas Wetlands in the Dinder National Park-Sudan. Environmental Processes:1-25. https://doi.org/10.1007/s40710-020-00444-5

## SUMMARY

Because of its rich biodiversity, the Dinder National Park, in Sudan, is recognized as a biosphere reserve, UNESCO World Heritage Natural Site and later has also been listed as a RAMSAR site in 2005. The seasonal and annual variability of the Dinder river flow have great impact on maya wetlands and hence on the habitats and the ecological status of the park. In addition, the Dinder river exhibits large morphological changes due to sediment transported within the river or from upper catchment, which affects inflows to mayas and floodplain inundation in general.

This chapter presents a quasi 3D modelling approach to support management of the valuable maya wetlands ecosystems, and in particular, assessment of hydrological and morphological regime of the Dinder river as well as the Musa maya. Six scenarios were developed and tested. The first three scenarios consider three different hydrologic conditions of average, wet and dry years of the existing system with constructed connection canal, and the other three scenarios consider the same hydrologic conditions but for the natural system without the connection canal. The modelling helps to understand the effect of human intervention (connection canal) on the Musa maya and shows the validity of the quasi 3D modeling approach to support decision making of maya wetlands management. The comparison between the simulated scenarios concludes that the hydrodynamics and sedimentology of the maya are driven by the two main factors: a) the hydrological variability of the Dinder river flow and b) sediment deposits on the inlet channel of the natural drainage. Without the connection canal, scenarios have shown that the maya is filled with water for wet year conditions only. However, with the existence of the connection canal, the maya can be filled with water for all hydrological conditions with water volumes estimated by 2.1, 2.4 and 1.3 million cubic meters for the average, wet and dry years, respectively. This would have beneficial impact for the maya ecosystem.

## 5.1 Introduction

Floodplain inundation and connectivity research in surface hydrology and geomorphology has experienced substantial evolution in the last decade (Heckmann et al., 2018). Previous studies concerned with hydrology (e.g. Bracken and Croke, 2007) and geomorphology (Brierley et al., 2006; Bracken et al., 2015; Heckmann et al., 2018), consider hydrological and sediment connectivity as a degree to which rivers facilitates the transfer of water and sediment into its floodplain.

It is well known that wetlands located in floodplains play a crucial role in maintaining the ecological functioning of the river ecosystem. They are characterized by high biodiversity, and hence, have attracted attention for preservation and restoration worldwide (Rebelo et al., 2012). The riverine wetlands reduce flood peaks and provide habitats for endangered species (Popescu et al., 2015). Ramsar Convention on wetlands recognizes wetlands as elements that need to be treated as part of the river system, and not as standalone units.

Floodplain wetlands are the most productive and valuable lands in terms of storage of flood water, groundwater recharge, retention of nutrients and unique habitat for wildlife (Li et al., 2019).

Fernandes et al. (2018) emphasized that in compound channels, the velocity gradient between the main channel and the floodplain flows leads to a flow structure more complex than in common single channels.

The role of flow and channel morphology in determining the structure of river ecosystems received little attention until the early 1980s (e.g. Nowell and Jumars, 1984). River rehabilitation and restoration requires good understanding and precise modelling. This includes the relationships between hydrological patterns, morphological processes and ecological responses in the river and its floodplain (Arthington et al., 2010).

Floodplain wetlands are existing all around the world. Previous studies on hydrological connectivity between river and its floodplain have considered such habitats as oxbow lakes (e.g. Gumiri and Iwakuma, 2002; Zeug and Winemiller, 2008; Glińska-Lewczuk, 2009), floodplain lakes/wetlands (e.g. Lew et al., 2016; Fernandes et al., 2018; Santisteban et al., 2019; Tan et al., 2019), seasonal wetlands (e.g. Yu et al., 2015; Li et al., 2019) and depression wetlands (De Steven and Toner, 2004; Cook and Hauer, 2007).

Assessment of the temporal and spatial morphological changes of floodplain and maya wetlands of the Dinder river in Sudan is very complex. Yet, it is very important to understand the hydrological processes and hence water supply for these important ecosystems. There are only few studies on watershed management and climate variability of the Dinder basin (e.g., AbdelHameed et al., 1997; Basheer et al., 2016), but the literature has no studies on the mechanism of the seasonal flooding of the maya wetlands.

The importance of the maya wetlands of the Dinder river is that it is a home of a very rich flora and fauna system in the Dinder National Park (DNP) in eastern part of Sudan close to its border with Ethiopia (please refer to section 2.2.2). Wetland inundation dynamics exert a strong control on processes such as plant productivity and water availability for wildlife animals during the dry season (AbdelHameed et al., 1997). Understanding inundation pattern of maya wetlands is important because it affects the vegetation pattern/type and it governs the life cycle of the biota. Flood inundation allows for movement of organisms and circulation of mineral substances, as well as enriching waters with dissolved oxygen, which enrich the nutrients required to feed the aquatic plants (Popescu et al., 2015).

The LULC changes in the Ethiopian highlands as reported by (Zeleke and Hurni, 2001; Bewket and Sterk, 2005; Hurni et al., 2005; Teferi et al., 2010; Teferi et al., 2013; Hassaballah et al., 2017) have apparently contributed to the existing high rate of soil erosion and land degradation in these areas (Bewket and Teferi, 2009). According to Ethiopian highland reclamation study (FAO 1984), the degraded area on the highlands estimated by 27 million ha of which, 14 million hectares is very seriously eroded with 2 million ha of this having reached a point of no return. As reported by Hawando (1997), the Ethiopian government has implemented enormous soil and water conservation activities through a "food for work" program. However, the magnitude of land degradation and the vastness of degraded land is so large that the impacts of conservation work seem comparatively small when viewed from a national perspective. Therefore, understanding the consequences of soil erosion in the Ethiopian highlands where runoff of Dinder river is generated, and integrating that into the sediment transport processes are very important for better management of the DNP ecosystems. The mayas feeders play a significant role in ecosystem conservation in view of the fact that it serves to route water and sediment across and out of the mayas.

During the last three decades, the DNP has experienced serious shortage of surface water, and hence limited green pasture during the dry seasons. This has directly affected the carrying capacity, particularly in Musa maya, which was chosen as a pilot maya for this study. The water shortage is claimed to be due to morphological changes and sediment deposition on the maya's feeder (locally known as the Saggai), which reduced the water flow from the river to the maya during floods. In addition, areas around the park are degraded as a result of mechanized farming and removal of the tree cover (Hassaballah et al., 2016). Although it is prohibited, grazing of livestock within the park causes competition with wildlife for pasture and for water.

The management plan of the DNP identifies a number of measures for conservation of the park, which focus on human activities such as control of poaching inside the park, control of mechanized agriculture around the park and adaptation of biosphere reserve concept. Ad hoc clearance of sediment from the feeders of mayas is also practiced to rescue wildlife in some years, which may not be sustainable. Only few measures have been implemented due to lack of funds (Mutasim and Frazer, 2004).

In addition to the human impacts, the main driving force for changes in the mayas system is the hydrological regime of the water supply system of the maya wetlands, particularly seasonal floods. Gomoiu (1998) reported that hydrological regime of a river system allows for exchange between aquatic and terrestrial habitats, ensuring that all components of the system are functioning. For effective management of the ecosystem, decision makers need to know which measures to implement for ecosystem rehabilitation and conservation and how these will improve the water system and the environment. There is no information on the hydrology or morphology of the Dinder river except of discharge measurement at station located just before the confluence with the Blue Nile. New field measurement within the DNP has been conducted as part of this study during 2013 to 2016.

This study presents an attempt to apply hydrodynamic and morphological modelling techniques to support decision making for the conservation of maya wetlands within the DNP. A quasi 3D numerical model using Delft3D software was built to simulate the behavior of the maya's system as a response to hydrological variability. Musa maya was chosen as a pilot maya for applying the hydrodynamic and morphology modelling approach.

### 5.1.1 Description of the pilot Musa maya

Musa maya (0.95 km$^2$) is situated about 10 km north-west of the Gelagu camp (Figure 5.1). In the past (around 1970's) it was one of the most productive mayas and an important source of water during the dry season. During the last three decades, the maya has experienced consecutive years of drought and observed to be dominated by grass that tolerant to drought. There is no distinct channel connecting the maya to the Dinder river. However, in June 2012, a connection canal was constructed to supply water to the maya after the consecutive years of drought.

## 5.2 Data and method

### 5.2.1 Collected data

A quasi 3D model was built to cover the river and the wetlands area, using Delft3D software. This model requires a topography grid map, surface roughness grid map, observed discharge data and cross-sections of the river Dinder. STRM (Shuttle Radar Topography Mission) DEM of 90 m resolution, produced by the National Aeronautics and Space Administration (NASA), was used. The data was expressed in geographic coordinates (latitude/longitude) and referenced to the WGS84. The model extent covered an area of about 105 km$^2$ inside the DNP (Figure 5.1). Since the vertical accuracy of the 90 m resolution DEM performs poorly in areas of moderate topographic variation and forestry area, a field topographic survey was conducted using an ordinary level and GPS to generate a DEM with higher accuracy for the model domain within the DNP.

***Water level data***

Due to absence of water level measurements within the pilot area inside the DNP, a Divers network was established in June 2013 to collect data for this study. The network consists of two Mini-Divers for continuous water column, temperature and atmospheric pressure measurement and one Baro-Diver to measure and correct for the atmospheric pressure changes. The divers were set to take measurements automatically on hourly basis. The divers were installed as follows: the first Mini-Diver was installed inside the Dinder river just upstream of the confluence of the river with the feeder of Musa maya to measure the water level of the river, and the second Mini-Diver was installed in the middle of the pilot Musa maya to measure the maya's water level. The Baro-Diver was installed near the other two Mini-Divers. This diver measures atmospheric pressure and is used to compensate for the variations in atmospheric pressure measured by the other two Mini-Divers. The locations of the water level monitoring points (WLMP) are shown in Figure 5.1.

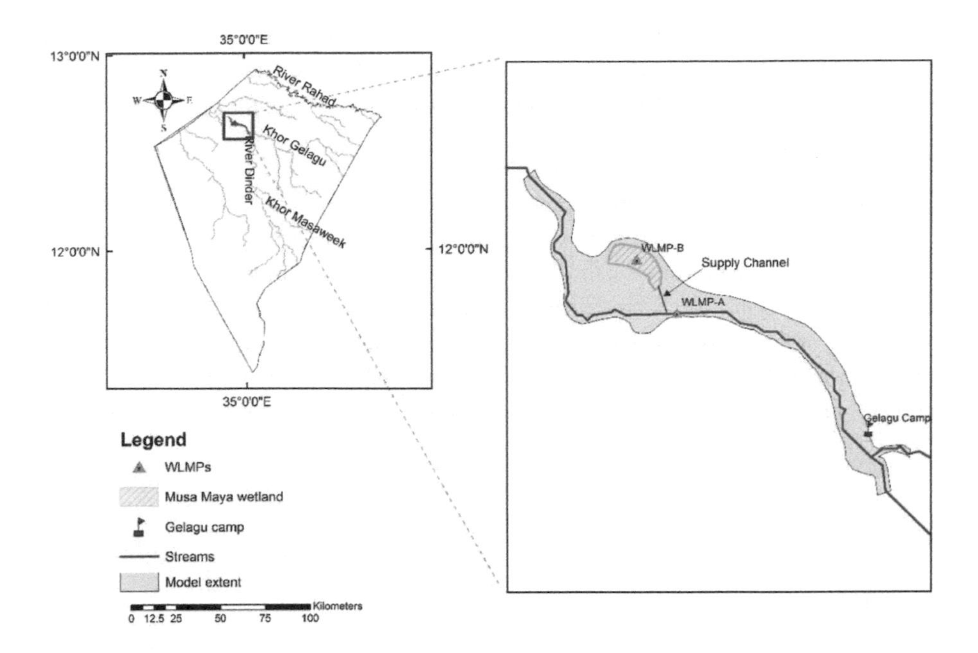


*Figure 5.1: DNP study area (left) and the model extent (right). Flow direction is from South East to North West.*

***River Cross sections***

During June 2013 (dry season), a topographic survey was conducted to measure river cross sections. The survey consisted of twenty-three cross sections, eighteen of them are in the Dinder river covering a reach of 20.0 km inside the DNP, three cross sections in Khor Gelagu and two in the main feeder of Musa maya. Each cross section was surveyed from above the highest flood level on the left bank of the river and crossing the river to a reasonable mark above the highest flood level on the right bank of the river. Ordinary land levelling was used for surveying banks and islands and a GPS for position fixing. The same cross-sections were repeated in 2016 to examine morphological changes.

***Sediment data***

Sediment data for the Dinder (suspended sediment concentration, without grain size analysis) are available for the years (1992-1995). The study site (within the morphological model domain) is completely inaccessible during the rainy season. Thus, suspended sediment data could not have been collected at the study site. The average suspended sediment concentrations were derived from the available daily data at Al-Gewisi station (1992-1995) and used for this study. Soil samples were taken from the pilot Musa maya and the Dinder river banks and bed during the field campaign in June 2016. Only suspended sediment can enter the maya. Therefore, the grain size distribution of the deposited sediment on the bed of the maya is assumed to be equal to the grain size distribution of the suspended sediment of the Dinder river.

***River channel data analysis***

River channel sinuosity in the study reach is low ($P \approx 1.28$). Analysis of the surveyed cross sections indicates that the river width within the study area varies between 180 m and 800 m with an average bank-full depth of 4 m during the survey period. The same cross sections measured in 2013 were repeated in 2016 to examine morphological changes. Comparison of the cross sections (Figure 5.2) shows that the changes vary from erosion to deposition with maximum bank erosion reaching 63 m on the right bank at cross-section 12 less than one kilometer downstream of the constructed connection canal.

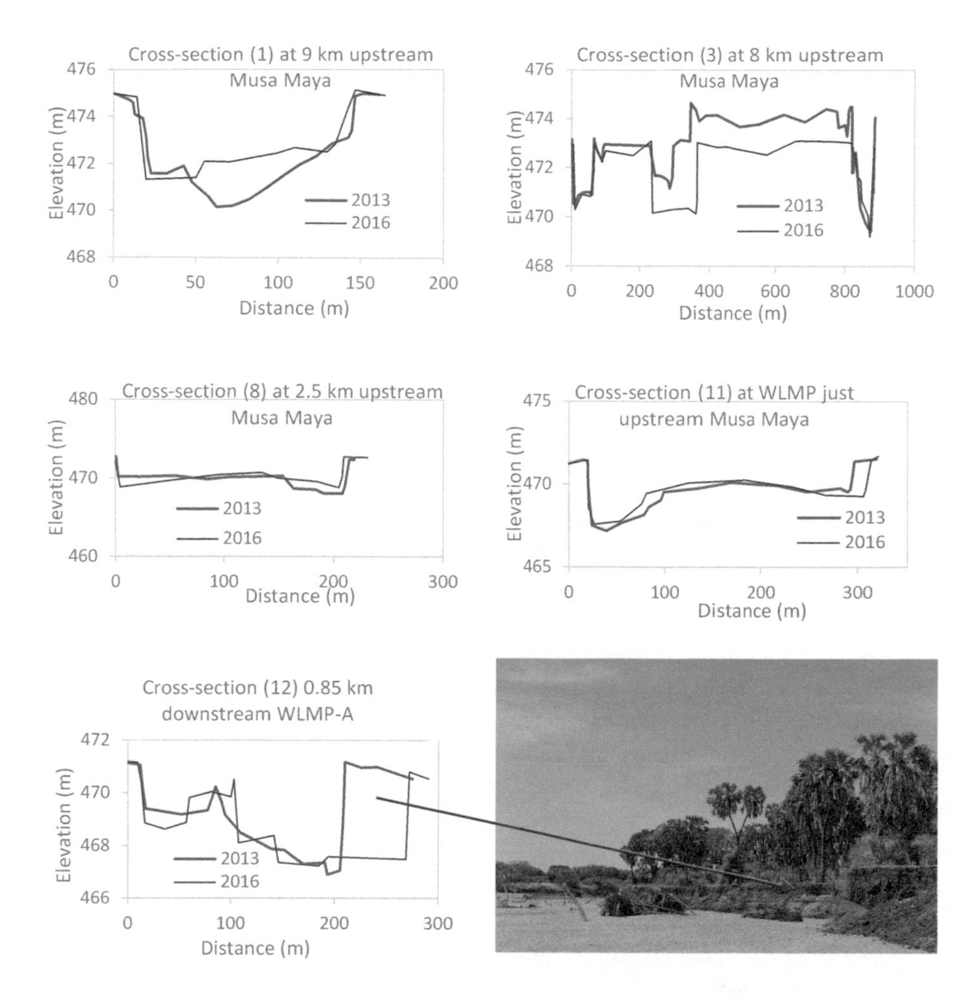


*Figure 5.2: Comparison of the surveyed cross sections between 2013 and 2016.*

Figure 5.3 shows that the water level in Musa maya has similar pattern to that of the Dinder river during the flood season. This indicates that the maya received its water from the river through the connection canal, and the contribution of direct rainfall and sheet flow from the surrounding area is very small.

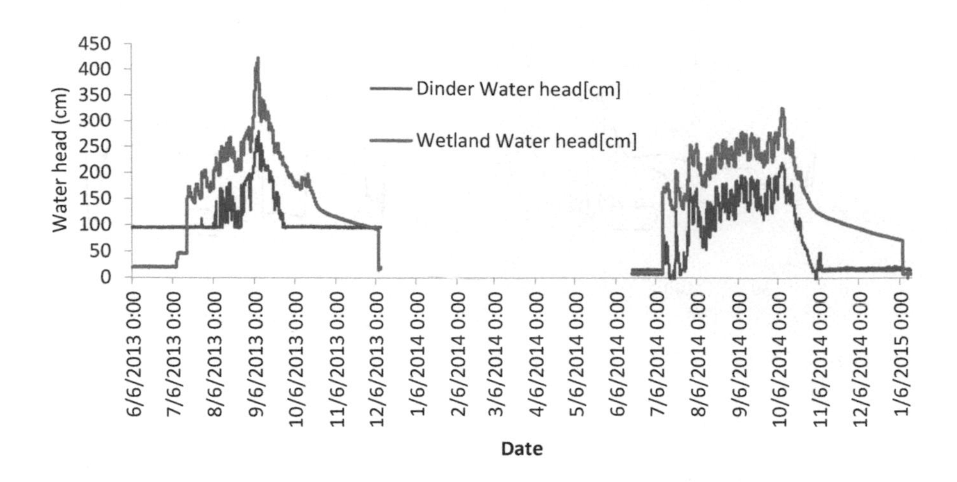


*Figure 5.3: Readings of the automatic gauges (a) at Dinder river and (b) at the pilot Musa maya. Red line is temperature (Celsius), green line is water pressure (cm $H_2O$) and blue line is the water column above Diver (cm).*

The analysis of the daily average sediment data for the period 1992 to 1995 showed that the suspended silt concentrations ranged between a maximum of 3339 mg/L at the beginning of the flood season (July) and a minimum of 140 mg/L in October (Figure 5.4).

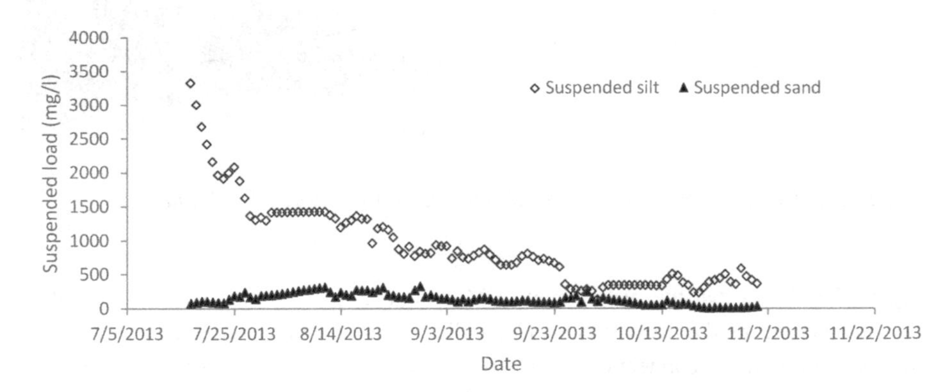


*Figure 5.4: The daily average suspended sediment concentration at Al-Gewisi station in the Dinder river.*

The analysis of the bed sample taken from the middle of the maya (during the survey of 2013) shows fine particles with a median diameter ($D_{50}$) of 22 μm. Silt is the dominant type of sediment in suspension, and it represents about 59% of the samples. Sand represents 32% and clay represent the remaining 9% of the suspended sediment materials (Figure 5.5).

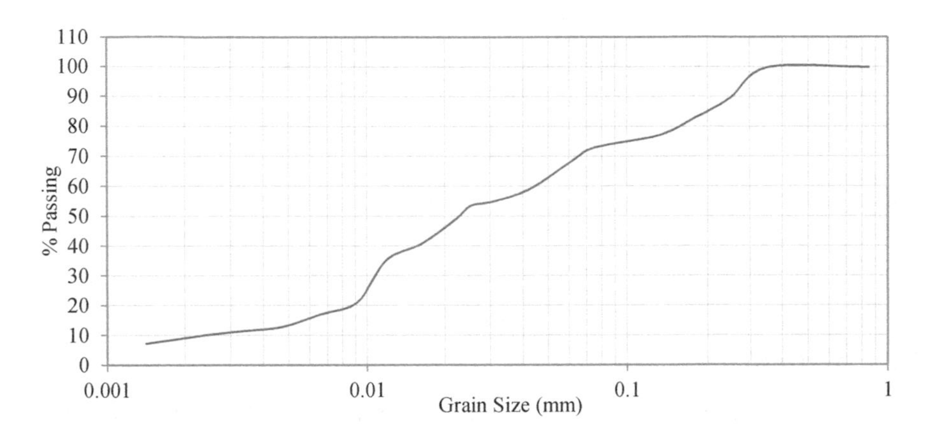


*Figure 5.5: Grain size distribution for Musa maya (bed sample)*

The analysis of the bed materials (Figure 5.6) at the upstream boundary shows that the sample has a $D_{50}$ of 420 μm decreased to 290 μm when analyzing sample at the downstream boundary of the model domain. Averaging results in $D_{50}$ of 355 μm for the model river reach. Thus, a $D_{50}$ of 355 μm was adopted for the Dinder river within the model domain. The analysis of bed sample taken from a location just downstream of the confluence of Khor Gelagu (tributary of the Dinder) shows coarse materials with a $D_{50}$ of 2600 μm. This can be attributed to the upstream contribution of coarser bed materials from Khor Gelagu tributary. Three samples were also taken from the river bank at the upstream boundary, downstream of the junction with Khor Gelagu and at the downstream boundary of the model domain. Analysis of the grain size distribution of these samples (Figure 5.6) shows $D_{50}$ of 57, 29 and 43 μm for the upstream boundary, middle part of the river reach and the downstream boundary, respectively resulting in an average $D_{50}$ of 43 μm.

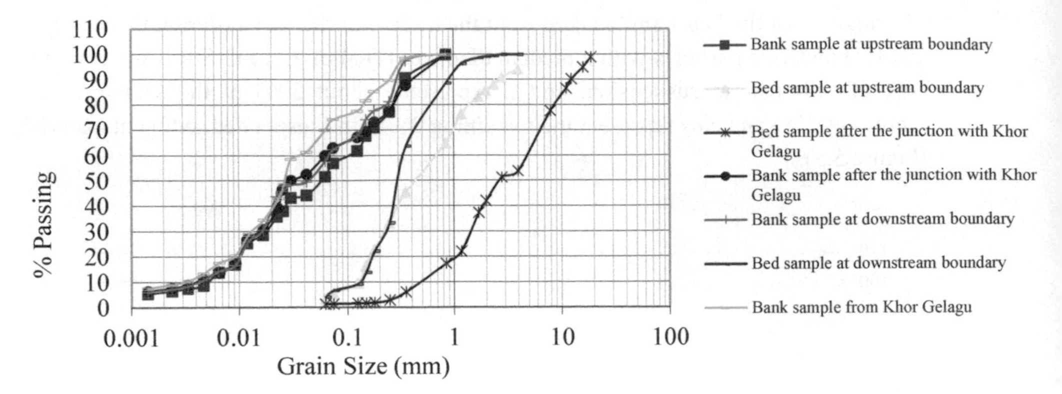


*Figure 5.6: Grain size distribution of the collected soil samples within the modelling area.*

## 5.2.2 Method

Since the objective of this study is to assess the hydrological and morphological regime of the Dinder and Musa maya, the hydrodynamic and morphological simulation is a plausible approach despite data limitation to at least understand the mechanism of water flow and sedimentation of the connection canal and Musa maya.

In this study, a quasi 3D morphological model of the Musa maya wetland was developed to understand the flooding mechanism and the morphological changes in the area. Bathymetry of the model is shown in Figure 5.7. The measured flow data at Al-Gewisi station were provided by the Ministry of Water Resources, Irrigation and Electricity, Sudan. The model has been set to a combination of grids, distributed over the area of the wetland (Figure 5.7). The grid cell size varies based on the topography and model domain with total number of computational cells of 47124.

Delft3D software was used to simulate both the hydrodynamics and the morphology of the Dinder river system and the sediment processes in its floodplain (maya wetlands). Delft3D software has been developed by Deltares (http://oss.deltares.nl/web/delft3d) to simulate hydraulic phenomena in river, estuarine and coastal areas. The software simulates variations in time and space (2D or 3D) of hydrodynamics, morphology water quality and sediment transport phenomena. Detailed description of the open-source code of the model is reported by Lesser et al. (2004). The overland flow and channel flow module in Delft3D that was set for the study area included a topography grid map, surface roughness grid map, observed discharge data and the surveyed cross-sections of the Dinder river.

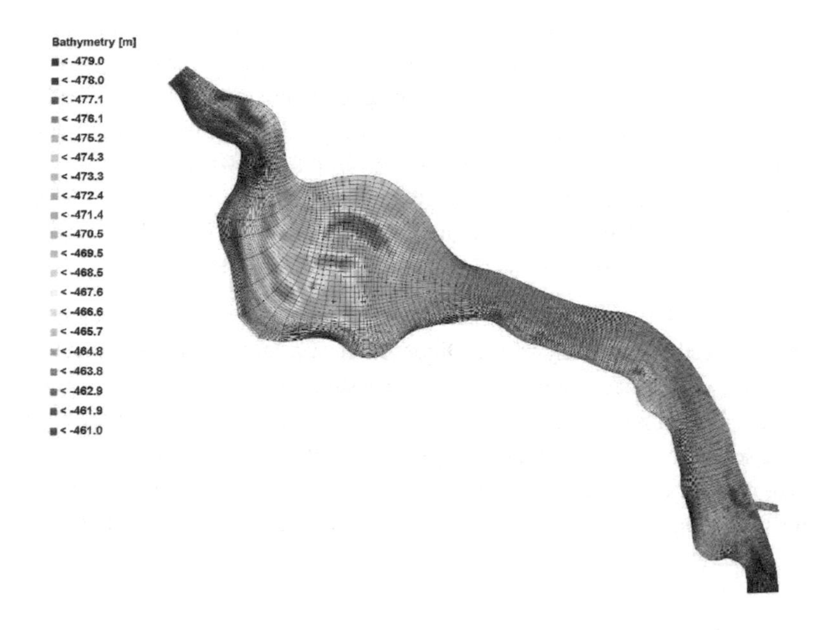


*Figure 5.7: Bathymetry of the study area, computational grid and bed elevations (masl) in 2013.*

***Setup of the hydrodynamic model***

We applied Delft3D to a 20 km reach of the Dinder river between Gelagu camp and up to few kilometers downstream of the pilot Musa maya (Figure 5.1). Topography data were obtained from the 90 m resolution DEM data from the Shuttle Radar Topography Mission (SRTM). Mukul et al. (2015) stated that SRTM data over Africa have an average absolute elevation accuracy of 16 m with 90% confidence (Root Mean Square Error of 9.73 m). The SRTM requires processing to remove vegetation effects so as to obtain a 'bare earth' DEM as the X- and C-band SRTM do not fully penetrate vegetation canopies (Wilson et al., 2007). To remove the vegetation effects, we used the ground survey data to correct the DEM in the targeted area. River channel characteristics (width, depth and slope) were approximated using the measured cross sections of the Dinder river reach upstream and downstream of the pilot Musa maya. We assumed that direct runoff and rainfall inputs to the maya balanced losses due to evapotranspiration and infiltration. Thus, detailed floodplain hydrologic processes were not considered. Three boundaries were set, two upstream boundaries of discharge inflow into the model, and the downstream boundary contained a discharge-stage relation. The water entered the Dinder system through the upstream boundaries that were set as river channels in the model. The model was run for the wet season of year 2013 with a computational time step of 12 seconds (0.2 minutes) starting from the dry condition. A constant Manning roughness coefficient n of 0.035, which corresponds to a normal river channel with some weeds and stones for the entire river reach was used. Al-Gewisi station approximately 130 km downstream of the DNP is the only hydrological station on the Dinder river. Therefore, the computed flow at a location just upstream of Gelagu camp through rainfall-runoff modelling (Hassaballah et al., 2017) was used as an upstream boundary condition for the model domain.

The downstream boundary condition was set as a Q-H relation which was calculated based on the cross-section geometry of the Dinder river at a location approximately 10 km downstream the gauge measuring station. The measured water levels (2013-2014) at two locations just upstream of the pilot maya and inside the maya were used for model calibration. The selection of the simulation time step depends on several parameters, such as the grid size of the model, the water depth, the required accuracy and the stability of the model during simulation. The Courant-Friedrichs-Lewy number (CFL) is defined as:

$$CFL = \frac{\Delta t\,\sqrt{gh}}{[\Delta x, \Delta y.]}$$

where $\Delta t$ is the time step (in second), $g$ is the acceleration of gravity, $h$ is the total water depth, and $(\Delta x, \Delta y)$ are the smallest grid spaces in $x$ and $y$ direction of the physical space. Generally, and for model stability, the $CFL$ should not exceed a value of ten (Deltares, 2010).

For the hydrodynamic model and the selected schematization of the grid cells, the time step used is 12 seconds and the value of $CFL$ varies in space and time. Other numerical parameters' values are coinciding with the default values of the Delft3D software. In the model setup phase, inaccuracies due to the large size of the computational grid cells were compensated for by manual adjustments of topographic levels, ensuring that the thalweg elevation in the model is close to the measured one.

***Calibration of the hydrodynamic model***

Estimation of model parameter values is difficult even with highly specialized laboratory experiments. A practical approach is to estimate such parameters from available process data or from the literature. Typically, only a subset of the parameters can be estimated due to restrictions imposed by the model structure, lack of measurements, and limited data.

In this study, the various model parameters were adopted from previous literature from the region (e.g. Omer et al. 2015). Then, manual calibration through adjusting the model topography and the Manning roughness parameter in an iterative way until the observed water level exhibited an acceptable level of agreement with model output.

Many simulations have been conducted using variable bed roughness until the modelled water level closely fitted the observed water level. This was obtained with a Manning roughness value of 0.035. Then, a systematic adjustment of the topography was made to improve the calibration fitting. The model results were compared with the water levels measured at the pilot Musa Maya and the Dinder River just upstream the pilot Maya.

A comparison was made between model output and water level data from the measured water level at the pilot Musa maya and the Dinder river just upstream the pilot maya (Figure 5.8). The Root Mean Square Error (RMSE) and the Nash-Sutcliffe efficiency (NSE) introduced by Nash and Sutcliffe (1970) were used to evaluate the model performance.

The NSE and RMSE ranged from 0.67 to 0.89 and 0.14 m to 0.30 m for the daily water level for the Dinder river and Musa maya respectively. The modelled water levels matched closely for the Dinder gauge and were under-predicted by a maximum value of 0.26 m at high water level. While, at Musa maya, the model is over-predicting the stage at low water levels and under-predicting the stage at high water level by a maximum value of 0.81 m. This is likely due to an error in the floodplain topography resulted from the combination of the DEM and the surveyed topography.

Despite the availability of digital elevation models plus extensive field surveys carried out in 2013 through 2016 in a very harsh forested environment, the topography of the Mayas wetlands floodplain might not be very accurate to model flood propagation. However, it is satisfactory to demonstrate how filling and emptying of Mayas wetlands occurs. For seasonal isolated Mayas, it is difficult to determine boundary conditions due to the complex flow patterns and limited in situ observations. Uncertainties in the hydrodynamic model might result in smaller defined Mayas wetlands boundaries, bathymetric variations and unconsidered groundwater recharge/discharge.

In general, the calibration results indicate that the Delft3D performance is reasonably good to understand the inundation mechanism of the maya. Then, the model was used to simulate the filling mechanism using variable river flows with average, wet and dry hydrological years and keeping other model parameters without change. Model validation with different data set is the second step after the calibration if another subset of data is available to perform this step. A substantial effort has been made to collect and gather data from a very harsh and inaccessible environment. These data for only two years were used for model calibration. Due to time and cost limitations, we were not able to repeat some of the measurements (e.g. water level and cross sections) needed for model validation.

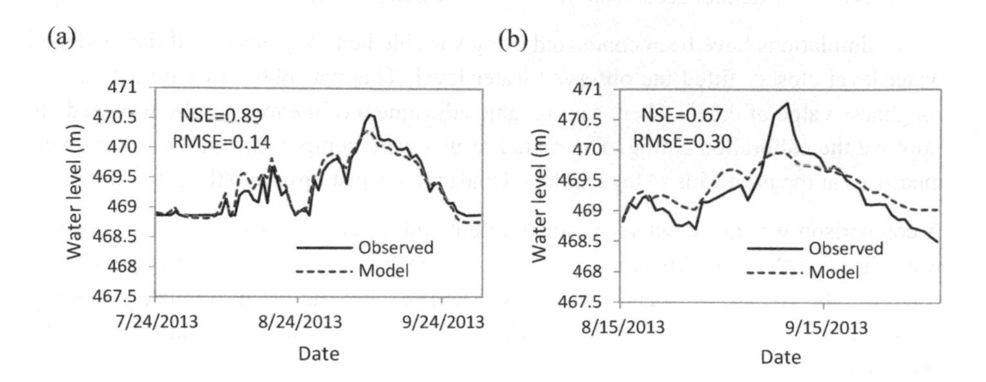


*Figure 5.8: Results of hydraulic model calibration at (a) Dinder river and (b) Musa maya: modelled vs. observed water levels.*

***Setup of the morphological model***

The calibrated hydrodynamic model is updated to include the morphological setup. The morphological model is based on the hydrodynamic model, with the aim to simulate sediment process inside the pilot maya and morphological changes in the river reach during the two flood seasons of 2013 and 2016. For non-cohesive sediment (sand), the Van Rijn (1984) transport formula was used to calculate the erosion and deposition. Whilst, for cohesive sediment fractions (silt) the fluxes between the water phase and the bed were calculated with the well-known Partheniades-Krone formulations (Partheniades, 1965). The model was calibrated on the measured cross-sections changes during these two periods derived from the field survey data. There were no data available on soil stratification. The morphological computations were excessively time-consuming due to the large number of computational cells. These large numbers of cells were unavoidable due to the complexity of the processes to be simulated, such as 2D hydrodynamics, bed change and suspended load transport, and sediment deposition.

The daily average sediment concentration of the river available for the period (1992-1995) is used as an upstream morphological boundary condition. The chosen values of the calibration parameters and the closure coefficients are given in Table 5.1.

*Table 5.1: Values of physical parameters derived during the calibration process*

| Physical parameter | Calibrated value |
|---|---|
| Spiral flow – ($\beta$) | 0.5 (-) |
| Horizontal eddy viscosity | 1.0 ($m^2$ $s^{-1}$) |
| Horizontal eddy diffusivity | 1.0 ($m^2$ $s^{-1}$) |
| Specific density of sediment | 2650 (kg $m^{-3}$) |
| $C_{soil}$ (reference density of hindering settling) | 1600 (kg $m^{-3}$) |
| $D_{50}$ | 355 μm |
| Dry density of sand | 2000 (kg $m^{-3}$) |
| Dry density of silt (deposited suspended solids) | 1200 (kg $m^{-3}$) |
| Ws,0 (settling velocity of suspended solids) | 0.005 (mm $s^{-1}$) |
| $\tau_c$ (critical shear stress for erosion of silt) | 1 (N $m^{-2}$) |
| $\tau_d$ (critical shear stress for deposition of suspended solids) | 1000 (N $m^{-2}$) |
| $M$ (erosion rate of deposited silt) | 2 (mg $m^{-2}$ $s^{-1}$) |
| Manning roughness *(n)* | 0.035 |

The critical bed shear stress for deposition of suspended solids is kept at 1000 N $m^{-2}$ which means the suspended sediment is always allowed to deposit. Below this value, any particle was free to deposit according to its fall velocity and depending on the computed bed shear stress. While, the critical shear stress for erosion is increased to 1 N $m^{-2}$ which means unless the bed shear stress exceeds this value, no bed erosion takes place; in addition, the erosion parameter rate is reduced to 2 mg $m^{-2}$ $s^{-1}$. These values are defined based on the calibration process trials.

## 5.3 RESULTS AND DISCUSSION

To understand the hydrological and morphological connectivity of the maya with the river, more specifically the filling/emptying process, six scenarios were investigated. The first three scenarios consider three different hydrologic-year types of wet (10% exceedance), average (50% exceedance) and dry (90% exceedance) for the existing system with the constructed connection canal. The other three scenarios consider the same hydrologic conditions but for the natural system without the connection canal. This should help to understand, the effect of intervention (e.g. connection canal) on the Musa maya. In, other words, it clearly explains the validity of the quasi 3D modeling approach to support decision making of mayas wetland management, in this complex natural system. The average, wet and dry years were defined based on flow data for the years 1900-2016.

### 5.3.1 Scenario 1: average hydrologic year for the existing system with the constructed connection canal

In this scenario, the system was simulated using the flow of the year 2013 as a representative for the average hydrologic years (50% exceedance), with annual flow in the range between 2.00 x $10^9$ $m^3/a$ and 3.00 x $10^9$ $m^3/a$, for the existing system with the constructed connection canal. The model results show that the constructed connection canal conveys and maintains the water flow from the river to the maya *"filling phase"*. This phase begins usually at the end of July when the water level in the main river at the cross-section just upstream of the connection canal rises to a level of 469 m (bed level of the connection canal). This phase continues until the water level reached its maximum levels during the first two weeks of September (Figure 5.9a and b). With such average flow, the maya is partially inundated with variable water depths according to the maya's topography with maximum depth of 3.5 m. Other parts of the floodplain remain dry with no connection with the river water. This phase is considered as the wet condition of the maya (Figure 5.9a and b). As the river water level falls, the "*drainage phase*" begins through the same connection canal and it continues until the water level is low enough that stagnant and dynamic waters are completely isolated (Figure 5.9c and d). The maya is left with an average water depth of 1.75 m and estimated water volume of 1.66 x $10^6$ $m^3$.

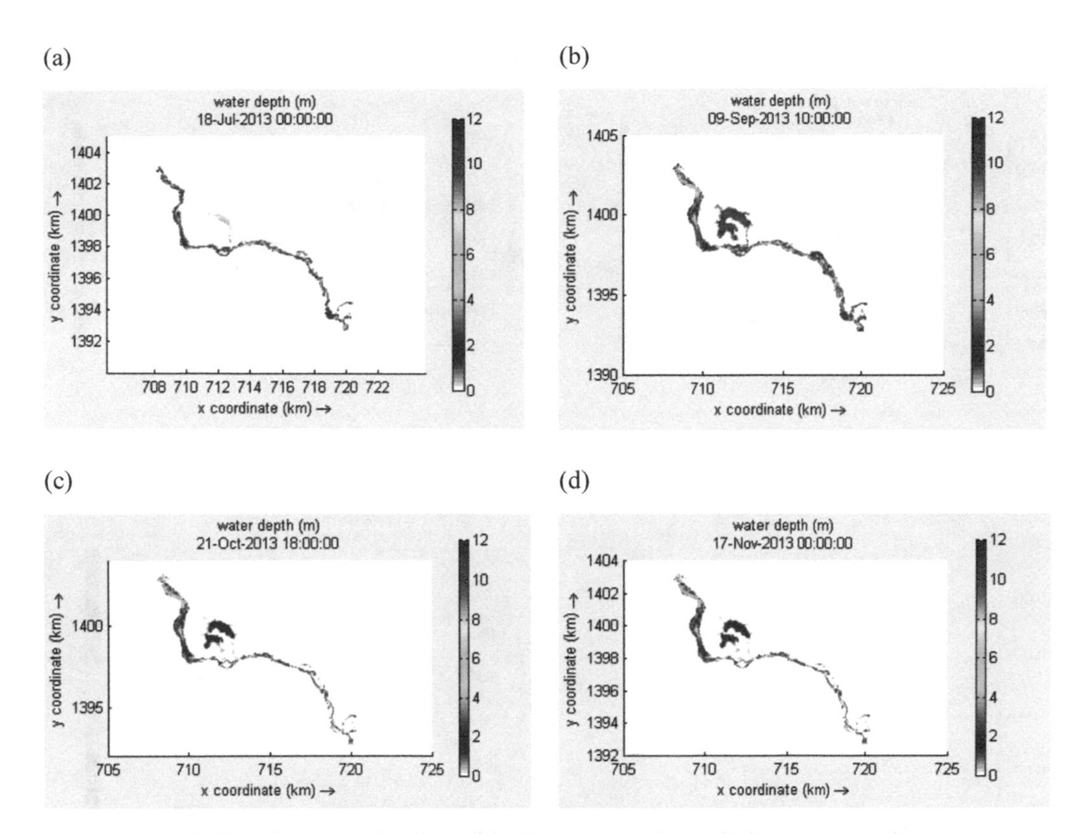


*Figure 5.9: Inundation mechanism of the Musa maya through the constructed connection canal considering average hydrologic year a) and b) filling phase, c) drainage phase and d) isolation phase.*

Low rate of sediment is transported into the Musa maya with flood and decreases gradually (Figure 5.10a and b). The sediment transport process continues at low rate until the maya and the river waters are completely disconnected. The remaining stagnant water in the maya remains with low suspended sediment concentration (Figure 5.10c). Within the river cross-section just downstream of the connection canal, high bank erosion on the right and deposition at the middle of the cross-section are observed (Figure 5.10d). No morphological changes are observed in the connection canal.

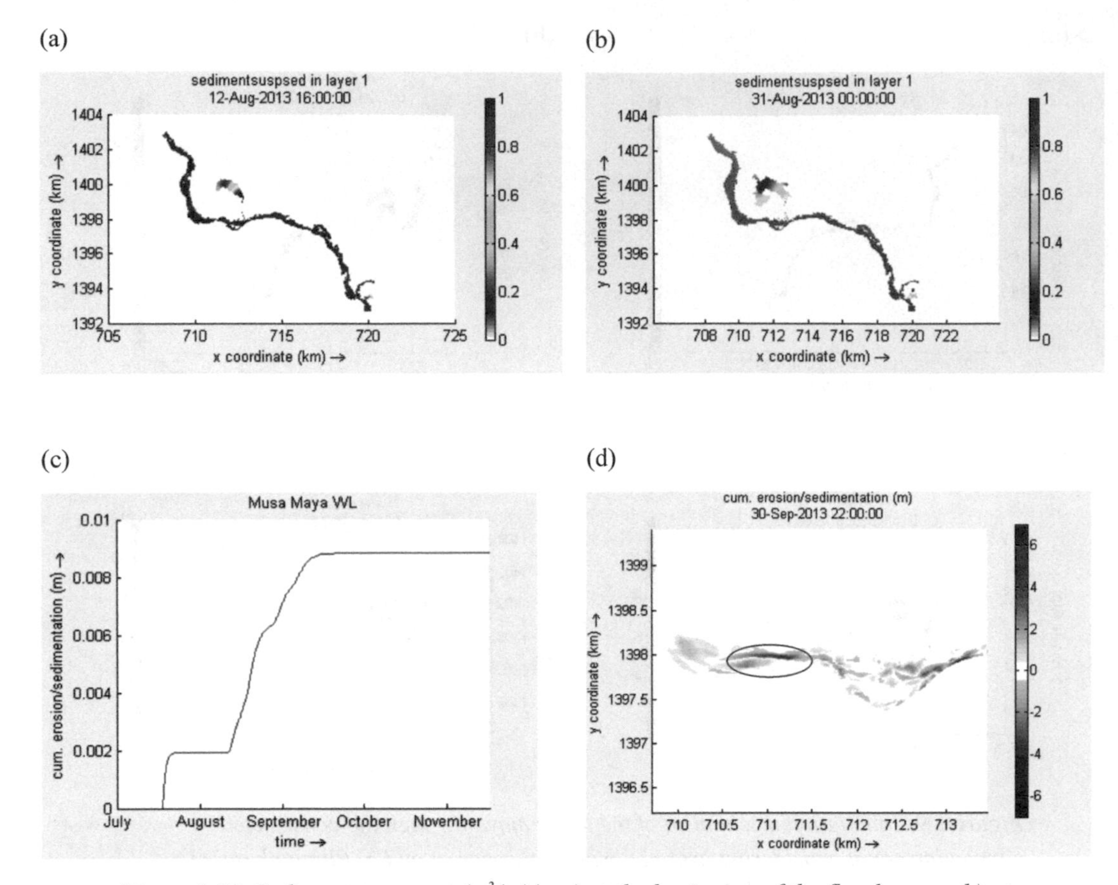


*Figure 5.10: Sediment transport ($m^3/m$)/s, a) at the beginning of the flood season b) at the end of the flood season and c) and d) show the sedimentation in the maya and along the connection canal.*

## 5.3.2 Scenario 2: Wet hydrologic year for the existing system with the constructed connection canal

In this scenario, the system was simulated using the flow of the year 2012 as a representative for the wet hydrologic years (10% exceedance), with annual flow equaled or exceeded 4.22 x $10^9$ $m^3$/a, for the existing system with the constructed connection canal. Similar to scenario 1, the results show that the constructed connection canal conveys the water flow and maintains the input of water from the river to the maya *"filling phase"* (Figure 5.11a). The large magnitude of the river peak flood, results in a complete inundation of the maya and flooding of the river floodplain (Figure 5.11b). The maya is inundated with variable water depths according to the maya's topography with maximum depth of 4 m. This phase is considered as the wet condition of the maya. As the river water level falls, the "*drainage phase*" begins through the same connection canal and it continues until the water level is low enough that lentic (stagnant) and lotic (dynamic) waters are completely isolated (Figure 5.11c and d). In this stage, the maya and the river waters are completely disconnected. In late November and early December, the river falls to its minimum water level until it runs down to its dry condition. The maya is left with an average water depth of about 2 m and estimated water volume of 1.90 x $10^6$ $m^3$. In addition to maya's inundation, the floodplain is also left with many small water ponds. This water is normally consumed by wild animals, evaporates or infiltrates to recharge groundwater.

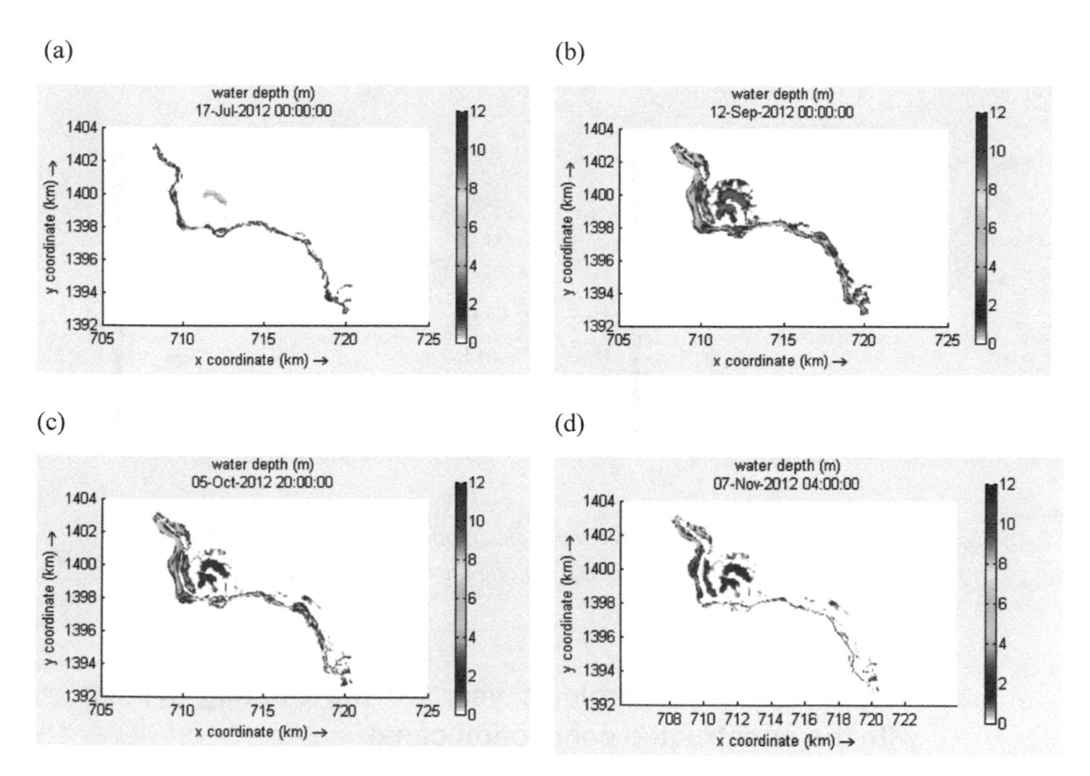


*Figure 5.11: Inundation mechanism of the Musa maya through the constructed connection canal considering wet hydrologic year a) and b) filling phase, c) drainage phase and d) isolation phase.*

Due to the construction of the connection canal, the model has shown that sediment is transported into and out of the Musa maya at a rate of 3 kg/m$^3$ during July and decreased gradually to less than 0.3 kg/m$^3$ in October. The sediment transport process was found to be accelerated during extreme floods compared to low floods. Within the river cross-section just downstream of the connection canal, very high bank erosion on the right and deposition at the middle of the cross-section are observed (Figure 5.12a). The result was the formation of an island. On the left bank, the river bed was silted up. Morphological dynamics within the connection canal itself is another important factor in the evolution of the maya and its aquatic components. Results show siltation process at the connection canal outlets. The deposition at the inlet of the connection canal and the formation of small delta at the tail of the connection canal at its connection with the maya is another evidence of the siltation process (Figure 5.12b).

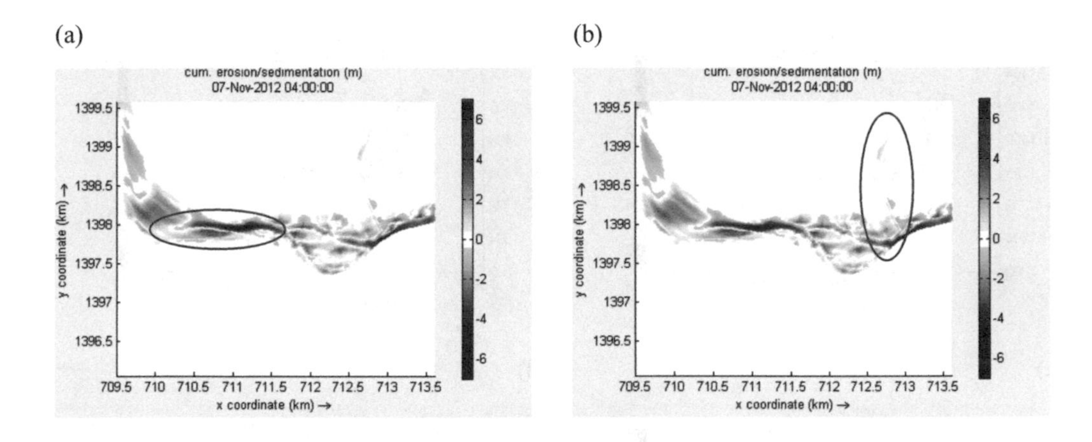


*Figure 5.12: Total cumulative erosion and sedimentation a) at the cross-section just downstream of the connection canal, and b) along the connection canal.*

## 5.3.3 Scenario 3: Dry hydrologic year for the existing system with the constructed connection canal

In this scenario, the system was simulated using the flow of the year 2015 as a representative for the dry hydrologic years (90% exceedance), with annual flow equaled to or less than 1.30 x $10^9$ m$^3$/a, for the existing system with the constructed connection canal. Same as scenario 1 and scenario 2, the result shows that the constructed connection canal conveys and maintains the water flow from the river to the maya *"filling phase"*.

But, with such small flow magnitude, the maya is inundated to shallow water depth with a maximum depth of about 2.3 m (Figure 5.13a and 5.13b). As the river water level falls, the "*drainage phase*" begins through the same connection canal and it continues until the water level is low enough that the stagnant and the dynamic waters are completely isolated. The maya left with an average water depth of 1.12 m and estimated water volume of 1.06 x $10^6$ $m^3$ (Figure 5.13c and 5.13d).

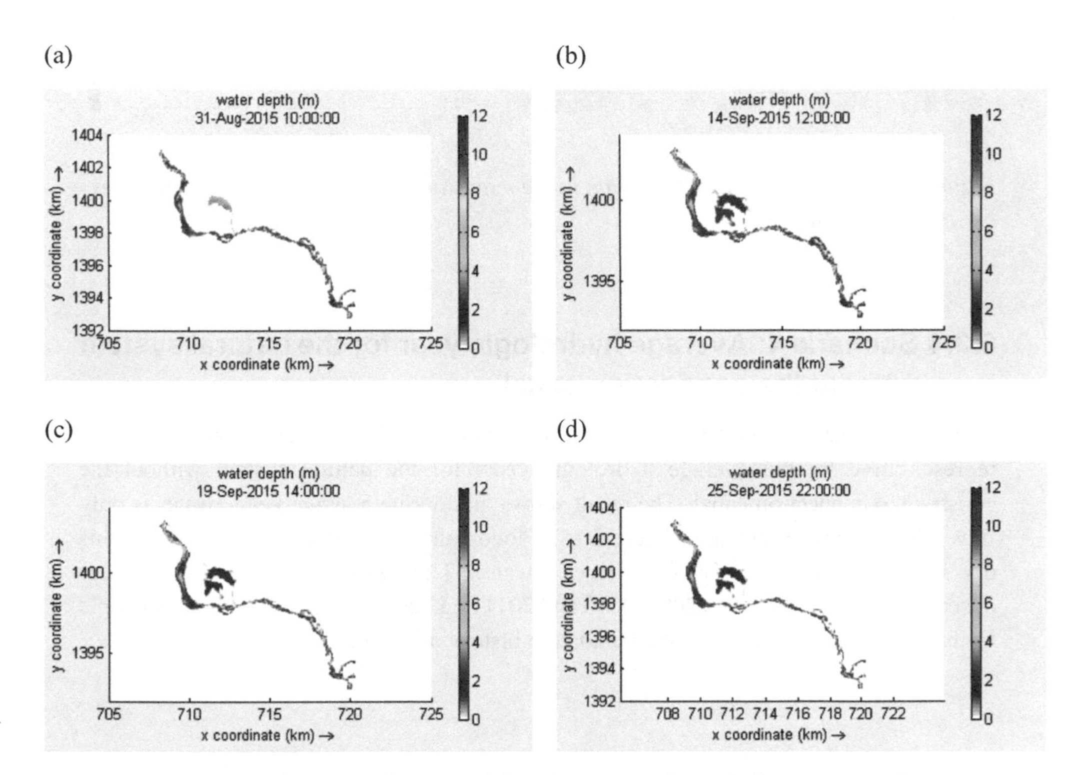


*Figure 5.13: Inundation mechanism of the Musa maya through the constructed connection canal considering dry hydrologic year a) and b) filling phase, c) drainage phase and d) isolation phase.*

Within the river cross-section just downstream of the connection canal, bank erosion on the right bank and deposition at mid cross-section were observed. Considering the small volume of water remains in the maya, Figure 5.14a has shown that the sediment deposition in the maya is very low (4 mm/a). No morphological changes were observed on the connection canal (Figure 5.14b).

(a) (b)

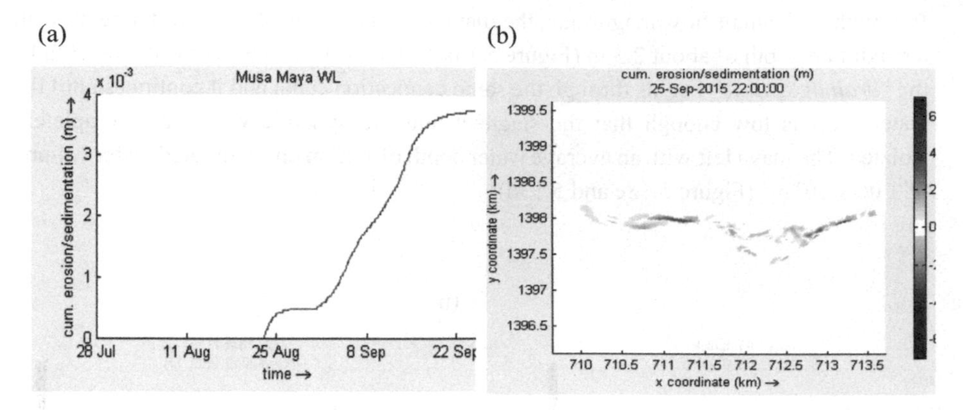


*Figure 5.14: a) the sediment deposition in the maya, and b) the morphological changes on the connection canal.*

## 5.3.4 Scenario 4: Average hydrologic year for the natural system without the connection canal

In this scenario, the system was simulated using the flow of the year 2013 as a representative for the average hydrologic years for the natural system without the constructed connection canal. The result shows that during average years, water is only flow within the river channel (Figure 5.15). Floodplain including the pilot maya remains dry with no connection with the river's water. This result was supported by our observation during the years 2009, 2010 and 2011 and our investigation with local people from wildlife police regarding the inundation history of the maya.

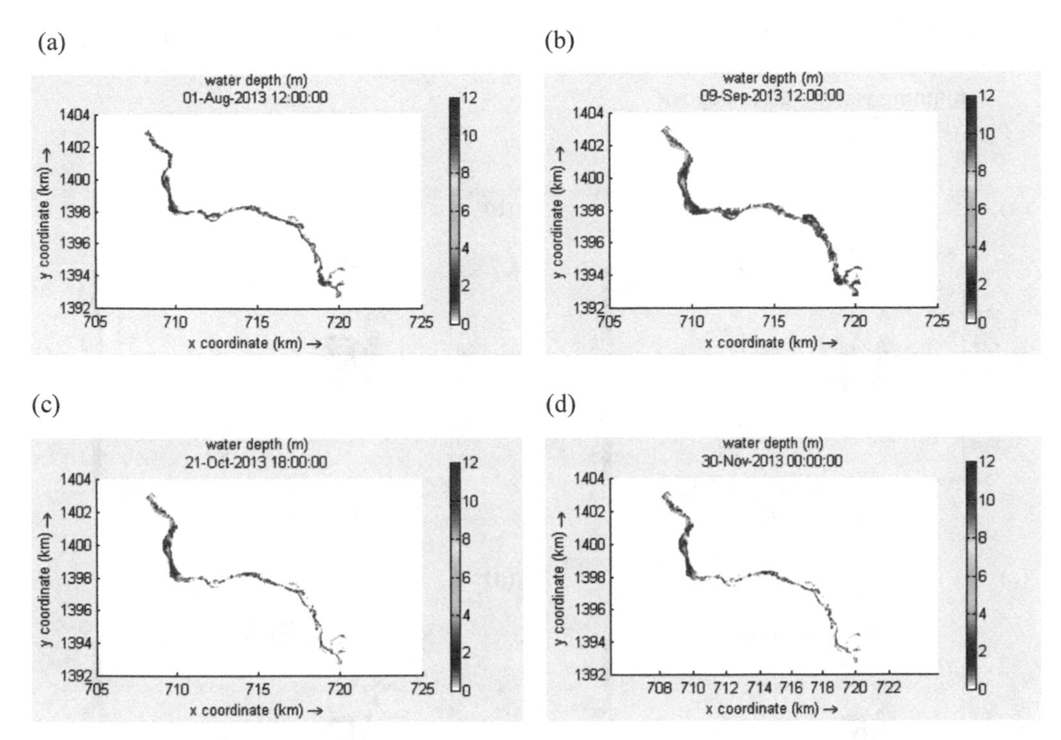


*Figure 5.15: During average hydrologic year, water is only flow within the river canal. Floodplain including the pilot maya remains dry.*

### 5.3.5 Scenario 5: Wet hydrologic year for the natural system without the connection canal

In this scenario, the system was simulated using the flow of the year 2012 as a representative for the wet hydrologic years for the natural system without the constructed connection canal. The result shows that when the river water overflows the bank-full level, floodplain is inundated through lowest area and the flood wave rises to a water level enough to inundate the maya "*filling phase*". The maya is completely inundated if the level in the river at the cross-section just upstream the maya rises to 471.75 m (Figure 5.16a and b). As the water level falls "*drainage phase*" begins and it continues until the water level is low enough that lentic and lotic waters are isolated "*isolation phase*". The drainage and isolation are occurring at the end of September and the floodplain remains with isolated maya and small water ponds (Figure 5.16c and d). The maya left with relatively high water-depth of about 2.7 m and estimated water volume of 2.57 x $10^6$ $m^3$.

In addition to maya's inundation, floodplain also left with many small water ponds. At this stage water in the maya becomes stagnant and the suspended sediment carried by the remaining water start to deposit.

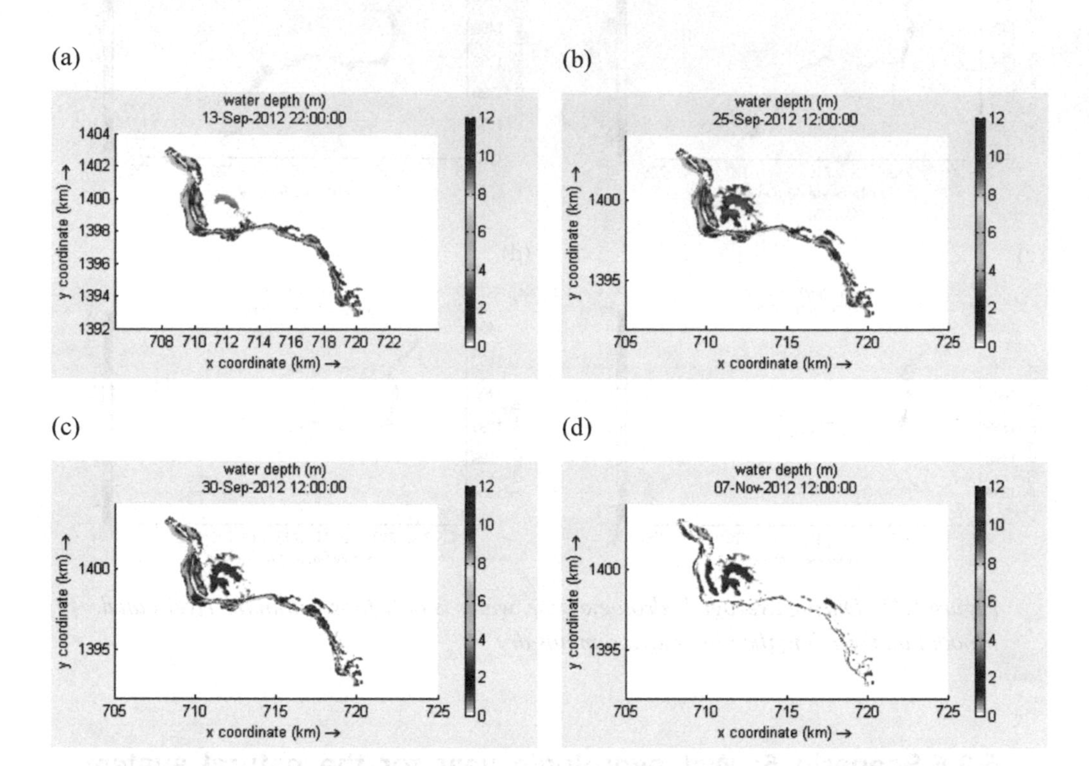


*Figure 5.16: Inundation mechanism of the Musa maya for natural system through overland flow a) and b) filling phase, c) drainage phase and d) isolation phase.*

The model has shown that sediment is transported into the Musa maya through overland flow during mid-September to late September (period of low sediment concentration) at low rate between 0.7 to 0.1 kg/m$^3$ (Figure 5.17a). By the end of the isolation phase, the stagnant water in the maya left with sediment concentration of less than 1 kg/m$^3$ (Figure 5.17b). Considering the low sediment concentration in water entering the maya, Figure 5.17c has shown that the sediment deposition in the maya is very low (less than 8 mm/a).

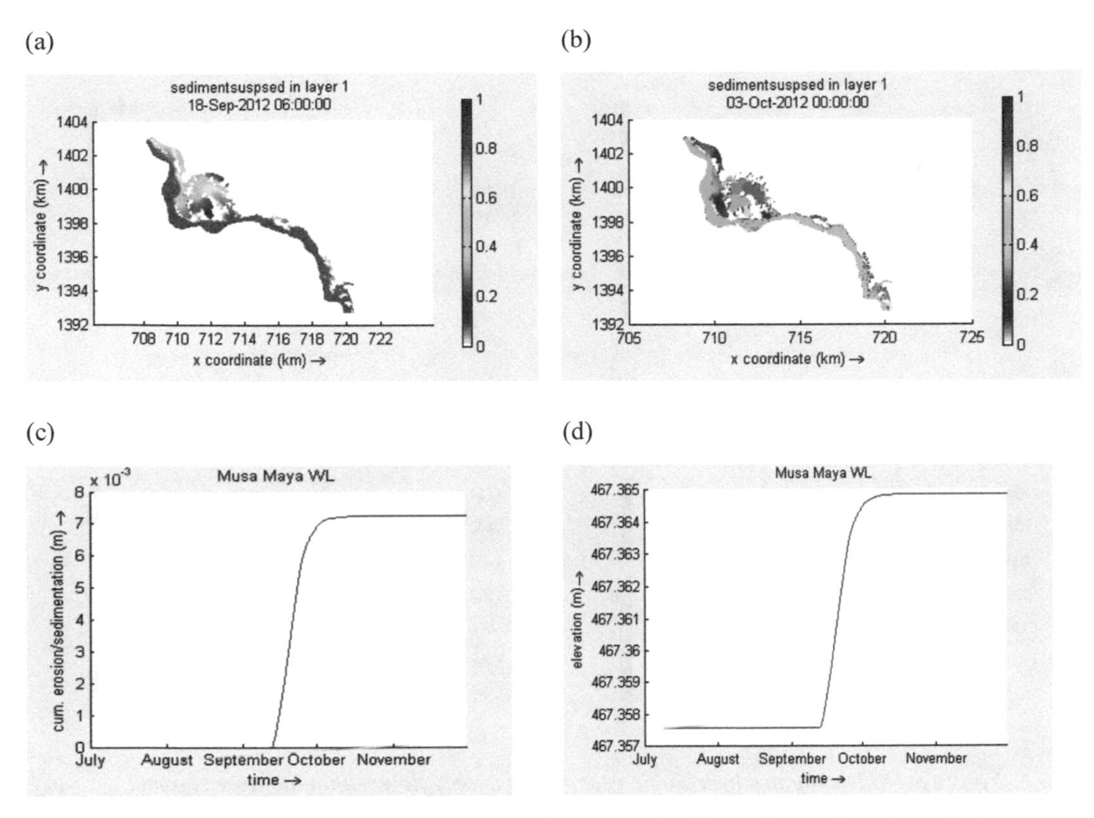


*Figure 5.17: Sediment transport into the maya, a) at the filling phase, b) at the isolation phase, and c) the deposition rate in the maya.*

## 5.3.6 Scenario 6: Dry hydrologic year for the natural system without the connection canal

In this scenario, the system was simulated using the flow of the year 2015 as a representative for the dry hydrologic years for the natural system without the constructed connection canal. Similar to scenario 5, the model result shows that during dry years the water is only flow within the river channel. Floodplain including the pilot maya remains dry with no connection with river's water (Figure 5.18). Summary of the six scenarios is presented in in Table 5.2.

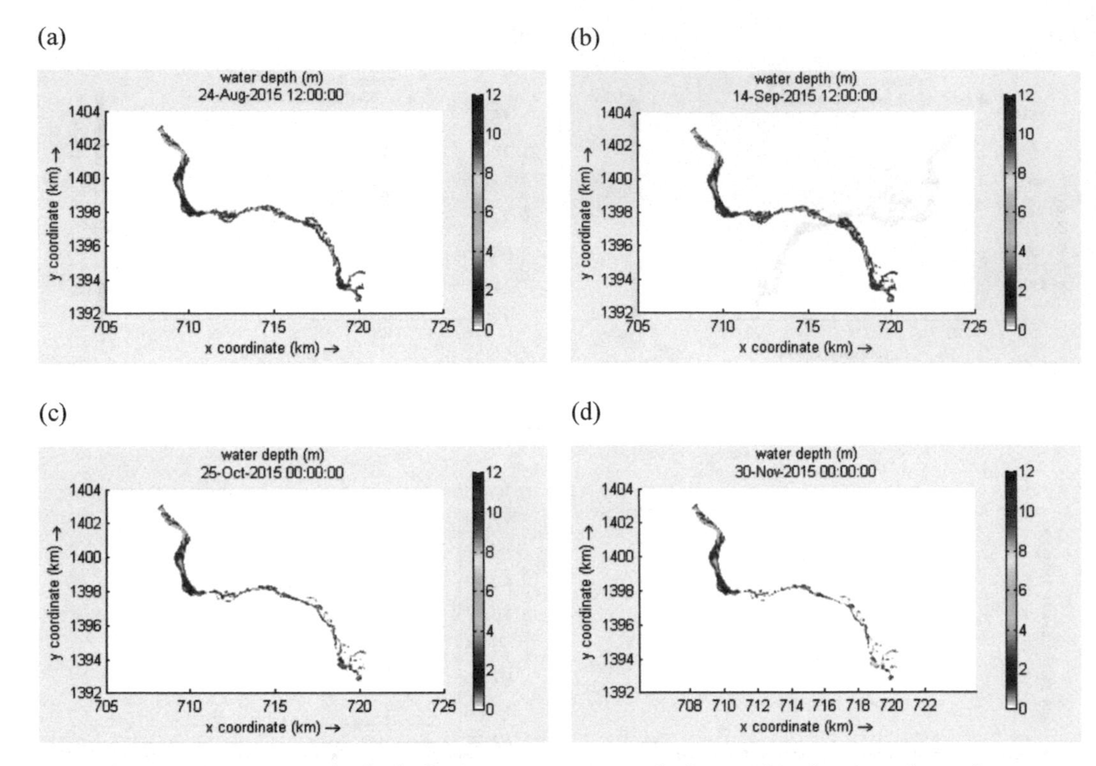


*Figure 5.18: During dry hydrologic year, water is only flow within the river channel. Floodplain including the pilot maya remains dry.*

*Table 5.2: Summary of the six scenarios; the first three scenarios consider three different hydrologic conditions of wet, average, and dry years for the existing system with the constructed connection canal (1-3). The other three scenarios consider the same hydrologic conditions but for the natural system without the connection canal (4-6).*

| Scenarios | Hydrologic year | Connection canal | Condition | Annual flow ($x10^9$ $m^3/a$) | Maya status | Maximum Inundation depth (m) | Inundation depth at the end of the flood (m) | Volume of water retained in the maya ($x10^6$ $m^3$) |
|---|---|---|---|---|---|---|---|---|
| 1 | 2013 | yes | average | 2.36 | inundated | 3.5 | 1.75 | 1.66 |
| 2 | 2012 | yes | wet | 5.59 | inundated | 4.0 | 2.0 | 1.90 |
| 3 | 2015 | yes | dry | 0.95 | inundated | 2.3 | 1.12 | 1.06 |
| 4 | 2013 | no | average | 2.36 | Not inundated | 0.0 | 0.0 | 0.0 |
| 5 | 2012 | no | wet | 5.59 | inundated | 4.0 | 2.70 | 2.57 |
| 6 | 2015 | no | dry | 0.95 | Not inundated | 0.0 | 0.0 | 0.0 |

## 5.4 Conclusions

This study presents the inundation mechanism and the morphological dynamics of the maya wetlands using a quasi 3D model. The study aims to improve our understanding of the filling mechanism of the maya wetlands and the effect of morphological change on filling of the mayas. SRTM data was used to generate the topography. However, since the vertical accuracy of the 90 m resolution DEM performs poorly in areas of moderate topographic variation and forestry area, a field topographic survey was conducted using an ordinary level and GPS to generate a DEM with higher accuracy with vertical error of 0.008 m and horizontal error of ± 3 m within the model domain. Channel topography was approximated using the measured cross-sections of the Dinder river reach upstream and downstream of the pilot maya. The approximation of the topography creates uncertainty to the bed topography. This is reflected by the morphological predictions. However, the comparison of the different scenarios suggests that:

- ✓ The hydrodynamics and sedimentology of the Musa maya are driven by the following two factors: a) the hydrological variability of the Dinder river flow, and b) sediment deposits on the bed channel of the natural drainage (feeder) that is likely caused by settled dead trees.
- ✓ Scenario 1: consider the existing system with the constructed connection canal for average years situations, the maya connected to the river only through the connection canal and partially inundated. Other parts of the floodplain remain dry with no connection with the river's water.
- ✓ Scenario 2: consider the existing system with constructed connection canal for high and extreme flood events. The floodplain depicts a high number of small water bodies with relatively long annual connection period (about a month) with the river's water. The Musa maya inundated to a maximum depth of about 4 m during the peak flood events. By the end of the flood season the maya left with an average water depth of 2.0 m.
- ✓ Scenario 3: consider the existing system with constructed connection canal for dry years, floodplain remains dry with no connection with the river's water. The maya connected to the river only through the connection canal for a short period and inundated to shallow depths.
- ✓ Scenario 5: consider the natural system without the constructed connection canal for wet years, the floodplain depicts a high number of small water bodies with short annual connection period (two weeks maximum) with the river's water, and the maya inundated to a maximum depth of about 4 m during the peak flood events. By the end of the flood season the maya left with an average water depth of about 2.7 m.
- ✓ Scenario 4 and 6: consider the natural system without the connection canal for average and dry years, water is only flow within the river channel. Floodplain including maya remains dry with no connection with the river's water.
- ✓ Three phases of hydrological connectivity were distinguished: 1) filling phase, 2) drainage phase, and 3) isolation phase.
- ✓ The modelling results have shown that the constructed connection canal has enhanced the filling mechanism of the maya, particularly in the dry years. However, the field measurement has shown a sever bank erosion (about 60m) on the right bank of the river cross-section just downstream of the connection canal.
- ✓ Modification of river width create more room for water within the river reach, and may lead to reduction in the water flows to the maya.

In conclusion, this chapter highlighted the value of integrating results from field observation and modelling in understanding the flooding and sediment transport processes in the maya wetlands and connected river sections. The development of a quasi 3D hydrodynamic model has improved our understanding of the hydrological functioning of the maya wetlands, allowing the quantification of the effects of possible future hydrological changes on mayas inundation. However, considering the high uncertainty in the data and model assumptions, beside unavailability of required data for model validation, the hydrological and morphological quantification should be considered as indicators for the effects of possible future hydrological and morphological changes, rather than exact values of the expected changes. Despite that, the model helped to understand the expected morphological changes of the maya's development.

The methodology applied in this study can be applied to other maya wetlands with similar hydrological conditions as the Musa maya. In particular, the maya wetlands that are located along the Dinder, such as Ein Elshamis and Greriesa mayas. The findings of the study present how maya wetlands are inundated and drain, and how the constructed connection canal has changed the hydrology of the Musa maya. Conservation of the maya wetlands ecosystem is crucial for sustainable development and utilization of the rich natural resources in the DNP. This issue requires integration of models that include both hydrodynamic and morphological models such as those presented in this study. As is often the case, the use of integrated modelling, besides the lack of reliable data introduces model uncertainties. Unfortunately, such unavoidable uncertainties are difficult to quantify in integrated modelling in general, and were not quantified in this study.

Furthermore, effective management of Dinder basin-wide, also requires international cooperation as the basin is shared between Ethiopia and Sudan. Strengthening of hydro-climatic monitoring networks is recommended to improve data availability and sharing for hydrological, hydrodynamics, morphological and other water management models to support decision-making processes. In this regard, the study recommends installation of continuous hydrological monitoring stations at the DNP. This should include water level measurements, flow and sediment measurements during flood season. Given the inaccessibility during the rainy season, automatic recorders are necessary in the context of the DNP. Though the present study looks at the maya wetlands flooding mechanism and sediment transport processes, it is important that this analysis is further expanded to include processes related to the ecological status of the maya wetlands ecosystem. Therefore, an assessment of the ecohydrology of the maya wetlands, in particular, the relations between vegetation dynamics and hydrological variability is presented in the next chapter.

# 6

# THE HYDROLOGICAL CONTROLS ON VEGETATION DYNAMICS AND WILDLIFE IN THE MAYAS WETLANDS OF THE DINDER NATIONAL PARK

[6] This chapter is based on: Hassaballah, K., McClain, M., Abdelhameed, S., Mohamed, Y., and Uhlenbrook, S.: The hydrological controls on vegetation dynamics and wildlife in the mayas wetlands of the Dinder National Park (submitted to Ecohydrology and Hydrobiology).

## SUMMARY

Relations between water availability, vegetation dynamics and wildlife were assessed in the maya wetlands of Dinder National Park (DNP) (Eastern Sudan). Field data on vegetation composition and wildlife were collected from four mayas to assess the ecosystem status and patterns of change. A systematic-random quadrat (SRQ) method was used to collect flora data. The normalized difference water index (NDWI) was used to estimate the inundation extent and the normalized difference vegetation index (NDVI) was used to estimate the related biomass in the mayas. Data on wildlife censuses were analyzed and relations to hydrological variability and vegetation cover were identified.

Seven plant species with different abundances were distinguished. A comparison between the mayas based on field observations revealed significant differences in both vegetation characteristics and density. The NDVI analysis of the data between 2001 and 2016 demonstrated significant variations in the area of vegetation cover. These variations were strongly linked to variations in the NDWI. The wildlife censuses showed that 84% of the total wildlife (herbivores) populations were found in the grassland within the periphery of mayas compared to only 16% in the burnt (human controlled) and open areas. This indicates that herbivores prefer grassland and woodland around the mayas rather than burnt and open areas. This is likely due to availability of water, pasture and shelter. Therefore, hydrological variability is found to be the key factor controlling the ecological processes.

## 6.1 INTRODUCTION

Many rivers in the world have suffered from a long history of degradation through direct and indirect human activities (Maddock, 1999). The magnitude and timing of water movements through river channels and floodplains have been changed by climate change, land use and land cover changes and morphological changes. The impacts of these changes from a conservation and ecosystem perspective have been widely documented (Maddock, 1999). Changes in physical structure of the river channel as well as its floodplain lead to changes in wetland surface area and subsequent degradation of ecosystem functions and services, which leads to changes in the composition of the biotic community inhabiting the river ecosystem, usually along with a reduction in the natural diversity of the ecosystem (Boon, 1992). If wetland area is lost, ecosystem services are also impacted. However, wetlands provide significant global ecosystem services such as biodiversity support, water quality improvement, flood attenuation and carbon management. Each of these services results from many physical-biological interactions.

Identification of the adverse consequences of both human and natural impacts on rivers, combined with an increase in overall environmental awareness, guided to many initiatives for river restoration as part of river basin management programs. Some river restoration studies intended to enhance the water quality (Jordan et al., 1990) while others intended to enhance the ecological integrity of river systems (RRP, 1993). No matter what the driving force are, there is a developing scientific knowledge related to theories, methods and effective applications of river ecosystem restoration being applied over the world (e.g. Brookes and Shields, 1996; Connelly and Knuth, 2002; Giller, 2005; Wohl et al., 2005; Kondolf, 2006; Palmer et al., 2010; Bernhardt and Palmer, 2011). Even though the recovery of lost biodiversity is challenged by invasive species, which thrive under disturbance and displace of natives. Not all damages to wetlands are reversible, but it is not always clear how much can be recovered through restoration (Zedler and Kercher, 2005).

Over the last decades, research on aquatic ecosystems associated with a direction towards river management and river restoration has increased. The role of streamflow and the river channel morphology in defining the structure of river ecosystems received little consideration until the early 1980s (Newbury, 1984; Nowell and Jumars, 1984). There are some important research goals relating to the spatial and temporal aspects of bio-physical habitat assessment. Maddock (1999) emphasized that upcoming studies on the growth of physical habitat assessments must attempt to integrate and combine the wide range of spatiotemporal scales that affect the ecosystem functioning and hence the human wellbeing.

The condition or health may be influenced by a number of factors relating to the river ecosystem, including its ecological status, water quality, hydrology, geomorphology and physical habitat. Both catchment hydrology and landscape ecology deal with processes and patterns and their interactions and functional consequences on different levels (Sivapalan, 2005; Turner, 2005). Accordingly, it is essential to study the relationship between hydrological and ecological processes to understand the two-way interactions.

The research on the interaction between hydrological and ecological systems relates to different levels and scales. A number of studies present an increasing linkages between hydrology and ecology in many research fields, such as ecohydrology (Richter et al., 1996; Wassen and Grootjans, 1996; Gurnell et al., 2000; Zalewski, 2002; Kundzewicz, 2003; Baird et al., 2004; Hannah et al., 2004) or riverine landscape ecology (Poole, 2002; Stanford, 2002; Tockner et al., 2002; Ward et al., 2002; Wiens, 2002; Schröder, 2006). With time, ecohydrology emerged as a new interdisciplinary field or even a paradigm (Bond, 2003; Hannah et al., 2004; Rodríguez-Iturbe and Porporato, 2005).

McCalin et al. (2012) underlined that ecohydrology is a trans-disciplinary science originated from the larger earth system science frameworks and examining common connections of the hydrological cycle and biological communities and is becoming a quickly developing branch of knowledge in hydrological science. It is likewise a connected science concentrated on critical thinking focused on problem solving and giving sound direction to basin-wide integrated land and water resources management.

As a summary from many papers in the field of ecohydrology, Zalewski et al. (2016) concluded that ecohydrology becomes an important bridge between ecology and environmental management at the catchment scale. As part of a basin-wide management, wetland ecosystem management is important to sustain the ecosystem integrity by protecting indigenous biodiversity and the ecological evolutionary processes that create and maintain that diversity. Challenged with the complexity inherent in natural ecosystems, achieving that goal will require that decision makers clearly describe anticipated ecosystem structure, role, and variability; illustrate differences between present and wanted conditions; define ecologically relevant and measurable indicators that can observe development toward ecosystem restoration, conservation and management goals (Noss, 1990; Cairns et al., 1993; Keddy et al., 1993; Dale and Beyeler, 2001; Carignan and Villard, 2002); and include adaptive strategies into resource management plans (Holling, 1978). If a wetland has continued filling and draining, its integrity is not necessarily preserved, nor is safe from future degradation. Wetlands degradation could be caused by hydrological alterations, sedimentation, salinization, eutrophication and exotic species invasions (Zedler and Kercher, 2005).

There are a large number of examples presenting the influence of hydrologic regime on ecological process and patterns and riverine landscapes (Schröder, 2006). As an example,

Naiman and Decamps (1997) along with Ward et al. (2002) assessed the ecological diversity of riverine landscapes. In such case, the changing environment support organism's adaptation to disrupted regimes over wide spatiotemporal scales (Lytle and Poff, 2004). Robinson et al. (2002) reported that the movement of many species is strongly linked to the spatiotemporal dynamics of the shifting landscape ecology. Tabacchi et al. (1998) assessed how vegetation dynamics are affected by the hydrological alterations and, on the other hand, how vegetation diversity and productivity influence riverine geomorphologic developments. Another example of research from the Nile swamps of southern Sudan by Petersen et al. (2007) reported that the swamp vegetation is water level dependent and that more swamp area becomes habitable for vegetation with decreasing water levels.

Similarly, there are many examples presenting the effects of ecological processes and patterns on hydrology. As an example, Tabacchi et al. (2000) analyzed the effects of riparian vegetation on hydrological processes, i.e.: (a) the effect of plant growth on water uptake, storage capacity and return to the atmosphere, (b) the control of runoff by the physical influence of living and dead plants on hydraulics, and (c) the effect of riparian vegetation functioning on water quality. Other well-known examples refer to so-called ecosystem engineers, i.e. organisms that are able to change environmental conditions through modifying, maintaining and/or creating habitats (Jones et al., 1994). A classic example of ecosystem engineering is a dam building beaver, that might have significant effects on hydrology, community structure and ecosystem functioning (Schröder, 2006; Nyssen et al., 2011).

The effect of flood-plain vegetation on hydrology and vice versa, is very clear in the case of the Nile swamps of South Sudan (Sudd) e.g. increasing flood level causes some plants to be dislodged from their place, floating downstream and cause channel blockages (Sutcliffe and Parks, 1999; Petersen et al., 2007).

To assess the condition of a wetland, a number of indicators are used. Weilhoefer (2011) indicated that indicators must be tangible and normally quantifiable measures, utilized as a part of numerous fields of research to demonstrate the state of a system and the major factors that put a system under pressure. In the situation of wetlands, indicators are used to assess the status of the ecosystem via observing trends in space and time and identify the causes of changes (Wardrop et al., 2007; Young and Ratto, 2009; Ockenden et al., 2012; Balica et al., 2013).

Satellite image processing provides tools for investigating ecosystem conditions using different methods and indices. Many indices were developed to highlight the important features of the image based on reflectance characteristics (Deep and Saklani, 2014). The normalized difference vegetation index (NDVI) is a common and widely used index, applied in research on global environmental and climatic change (Bhandari et al., 2012).

Many studies successfully delineated open water bodies including wetlands from Landsat MSS, TM, and ETM+ images using the normalized difference water index (NDWI) technique (McFeeters, 1996; Jain et al., 2005; Sethre et al., 2005; Xu, 2006).

The Dinder river is the main source of water for the diverse ecosystem of the DNP and maya wetlands (Hassaballah et al., 2016). "maya" is a local name for floodplain wetlands found on both sides along the Dinder river. Mayas are oxbows cut-off from the meandering river. During recent years, the Dinder river has experienced significant changes in floodplain hydrology (i.e. dryness of some mayas), and the causes are not fully understood. This very likely has significant consequences on the maya ecosystem functions and services, but the associated processes, effects and dependencies need to be understood better. In previous studies we provided a detailed description of the mayas of the DNP (Hassaballah et al., 2016), the long-term trends in hydro-climatology of the Dinder and Rahad basins (Hassaballah et al., 2019), analysis of streamflow response to land use and land cover changes using satellite data and hydrological modelling in the Dinder and Rahad (Hassaballah et al., 2017), and modelling the hydrodynamic and morphology of the mayas. It is important that this analysis is further expanded to include processes and functions associated with the ecological condition of the mayas ecosystems. Therefore, this chapter aims to gain further insights about the ecohydrology of mayas wetlands of the DNP and to assess the relations between vegetation dynamics, hydrological variability and wildlife population size and structure.

## 6.2 METHODS

Field data on vegetation composition and wildlife were collected from four mayas. The vegetation data was collected using a systematic-random quadrat (SRQ) method and aims to assess the vegetation dynamic and patterns of changes. Data on wildlife was obtained through wildlife censuses. Then the normalized difference water index (NDWI) was used to estimate the inundation extent and the normalized difference vegetation index (NDVI) was used to estimate the related biomass in the maya. Finally, the wildlife data were analyzed and relations to hydrological variability and vegetation cover were identified. Musa maya was selected as a pilot maya for assessing the interlinkage between hydrology and flora & fauna.

### 6.2.1 Quadrat sampling for plant species

In this study, a systematic-random quadrat (SRQ) method was used by a Quadrate (40 $cm^2$) for collecting information regarding flora inside the mayas through four transects. The position of the first quadrat was chosen randomly which automatically determined the positions of all other quadrats in the sampling (25 m spacing). For instance, a quadrat is placed on the ground at random to count the vegetation within the sample. This was done to identify a) the type and category of maya (productive or not, young, mature, old), and b) the type and status of flora including distribution and density.

The field survey which conducted in June 2016 (dry season) included four mayas which represent three drainage systems. Ras Amir and Abdel Ghani represent the Khor Gelagu system. The Eastern bank of Dinder river drainage system is represented by Musa maya, while the Western bank is represented by Gererrisa maya. The objective was to collect a dataset that describes the ecological parameters of these mayas as related to the major biophysical features, including observations on their water regime and floral composition. Such data will reflect if the maya's condition enhances ecological services and values.

## 6.2.2 Delineation of water and vegetation surfaces

The NDWI and NDVI methods are used to extract the water and vegetation features presented in the satellite images of Musa maya. Musa maya suffers from drought for many years, and thus has been chosen as a pilot maya to apply these methods. Vegetation cover is one of the most important biophysical indicators that describes the ecological parameters, including the floral composition, which can be estimated using vegetation indices derived from the satellite images. NDVI and NDWI are calculated as:

$$NDVI = \frac{NIR - RED}{NIR + RED} \quad (1)$$

$$NDWI = \frac{GREEN - NIR}{GREEN + NIR} \quad (2)$$

Where RED is visible red reflectance, and NIR is near infrared reflectance. The wavelength range of NIR band is (750-1300 nm), RED band is (600-700 nm), and GREEN band is (550 nm).

Satellite images at a 30 m resolution of the Landsat TM, ETM+, and OLI images from Landsat 5, Landsat 7, and Landsat 8 were obtained from the Landsat archive (http://landsat.gsfc.nasa.gov/) for the years 2002, 2003, 2012, 2013, 2014, 2015 and 2016 (Row 171/Path 51). These data were chosen in different seasons after different hydrological conditions (average, wet and dry hydrological years). The years 2002, 2012, 2013 and 2015 were chosen for the NDWI analysis and the years after each (i.e. 2003,

2013, 2014 and 2016) were chosen for the NDVI analysis. Table 6.1 shows the hydrologic condition of the selected years for NDWI analysis. For the purpose of NDWI calculations, images were selected during the flood season each September to ensure capturing the maximum flood extent. While for NDVI analysis, images were selected during the dry season each May to quantify the vegetation coverage within the maya. Since all Landsat 7 ETM+ images collected after May 31st 2003, when the Scan Line Corrector failed, have data gaps, gap-filling of these images was processed using the layer stack tool in ERDAS Imagine 9.2 following the procedure proposed by USGS at https://landsat.usgs.gov/gap-filling-landsat-7-slc-single-scenes-using-erdas-imagine-TM.

*Table 6.1: Years selected for NDWI analysis and their hydrologic conditions (for Musa maya).*

| Years for NDWI | Annual flow ($10^9$ $m^3/a$) | Hydrologic condition | Years for NDVI |
|---|---|---|---|
| 2002 | 0.54 | Dry (with natural system) | 2003 |
| 2012 | 5.59 | Wet (with constructed connection canal) | 2013 |
| 2013 | 2.36 | Average (with constructed connection canal) | 2014 |
| 2015 | 0.95 | Dry (with constructed connection canal) | 2016 |

## 6.2.3 Wildlife assessment

A wildlife assessment requires a lot of field work, but provides important data for wildlife protection and conservation. Proper planning, implementation and evaluation of conservation programs depend, among other things, on the size and distribution of wildlife populations. This is particularly true for species which are under high threat from poaching or habitat loss. Many methods were designed for wildlife census purposes and they often vary in accuracy and details, (e.g. satellite and aircraft census, on ground direct counting and from the effects and residues left by animals (Mubarak, 2010). In this study we have analyzed data obtained from two wildlife censuses conducted in the DNP in 2010 and 2016 (Mubarak, 2010).

The census for the wildlife in the DNP from 11-12 May, 2010 was carried out. This census was done jointly by the General Directorate of Wildlife and Wildlife Research Center under the supervision of the Wildlife Research Station in Dinder. The method of counting wildlife in this census was based on direct counting using binoculars and telescopes (Mubarak, 2010). The census starts in the early morning and ends in the evening. This census assumes that the animals are naturally or homogenously spread over the areas covered by all three ecosystems in the DNP.

Census data of Warthog *(Phacochoerus aethiopicus)* in the DNP for the year 2016 were obtained from the census research conducted by Hassan (2017). The census was carried out during the February-May dry season. The aim was to estimate the population size and structure of Warthog and to determine its habitat preference. In this study we analyzed census data in four transects, namely Ras Amir maya, Musa maya, Gererrisa maya, and Abdelghani maya. Each transect begins from Gelagu the main camp inside the DNP and ends at one of the above mentioned mayas (Hassan, 2017).

## 6.3 RESULTS AND DISCUSSION

### 6.3.1 Quadrat sampling results

The results of plant composition (Table 6.2) in the four surveyed mayas showed seven plant species with different abundances.

*Table 6.2: Plant composition per maya.*

| | Plant spp* | Local name | Percentage of Plant composition per maya | | | |
|---|---|---|---|---|---|---|
| | | | Abdel Ghani | Musa | Ras Amir | Gererrisa |
| 1 | *Vossia cuspidata* | Hileiw | **27** | 0 | 0.5 | **30** |
| 2 | *Ipomoea acquatica* | Arkala | **68** | **29** | 2 | **56** |
| 3 | *Boerhavia coccinea* | Tirba | 3 | 5 | 1 | 7 |
| 4 | *Zornia glochidiata* | Shelni Maak/Luseig | 2 | 0 | 0 | 0 |
| 5 | *Cyperus sp* | Seida | 0 | 1 | **36.5** | 7 |
| 6 | *Digera muricata* | Lublb | 0 | **65** | 0 | 0 |
| 7 | *Cassia obtusifolia* | Soreib | 0 | 0 | **60** | 0 |

*(*spp. stands for species pluralis, Latin for multiple species).*

***Ras Amir maya***

This maya is located about 13 km north east of Gelagu Camp at Long 12° 36' 51.4" N and Lat 35° 05' 48.4" E. Lake Ras Amir is the largest maya which occupied about 4.5 km$^2$. Its feeder has two branches which meet before entering the maya and feed it only when there is an excessive flood. To the north-east of Ras Amir, there is a large catchment area with many shallow gullies that feed the maya through runoff. The maya's bed was observed with no vegetation during 2001, except for a few herbs and scattered shrubs. Although in 1999 and 2000 it was full of water, in 2001 it was dry, but water was pumped to the maya to maintain the ecosystem functioning.

During June 2016 (end of the dry season) survey, the maya was full of water with many grasses and herbs. *Cassia obtusifolia* has the highest coverage (about 60 %). It is a plant of water-edges, floodplains, drainage, woodlands and grasslands in wetter tropical and subtropical areas. This plant has a range of tolerance both to climate and soil type, and it is invasive in parts of eastern Africa and observed to be toxic for animals when large quantities are eaten (Dunlop, 2007).

*Cyperus sp* is the second dominant type of grass observed within the maya. This is an aquatic annual or perennial plant species growing in stagnant or slow-moving water up to 0.5 m deep. *Cyperus* species are eaten by the larvae of some insects that include butterflies and moths. The tubes and seeds are an important food for many mammals and birds. The existence of these two plants species in Ras Amir maya indicates that this maya used to be inundated very frequently and hold water up to the next flood season. This was confirmed by the wildlife personnel in the DNP, and through our direct observation during our yearly visits to the DNP from 2011 to 2016.

***Musa maya***

This maya is located at the East-bank of the Dinder river at Long 12° 39' 25.5" N and Lat 34° 57' 08.6" E, with a total area of about 0.95 $km^2$ at a distance around 10 km west of Gelagu Camp. It was dry during the ecological base-line survey of 2001 being surrounded with *Sorghum Sudanese* (*S. Sudanese*), commonly known as Sudan grass. Sudan grass is tolerant of drought and warm temperatures. It has many roots which makes it one of the competent grasses for water absorption with small leaf area to reduce the evapotranspiration. The potential hazard from this plant is that the leaves at the early growing stage may contain varying amounts of cyanide, that separates and discharges a toxic substance known as prussic acid or hydro-cyanide when eaten (Armah-Agyeman et al., 2002). Cyanide compounds form if the plant is under stress (e.g. drought). At the point when wildlife eat plants that are high in this poison, they can die (Armah-Agyeman et al., 2002).

During the June 2016 survey which comes after the dry hydrological year of 2015, the maya was relatively dry with a small amount of water in the middle of the maya. *Ocimum basilicum* (Reihan) plants and greenery spread *of Corchorus spp* (Mulokheyia) are on the periphery of the maya, and *Digera muricata* is the dominant plant that covered most of the maya's area (about 65%). *Digera muricata* is a yearly 20 to 70 cm tall herb. A very adaptable plant that is seen growing naturally in both tropical and subtropical areas, where it can be found in semi-arid through to wet areas (Fenner, 1982). This plant most likely plays a crucial role in preventing soil erosion because it can provide a closed cover of vegetation on bare soil in a very short time.

***Abdelghani maya***

This maya is located about 2.0 km northwest of Gelagu Camp at Long 12° 36' 40.5" N and Lat 35° 01' 38.6" E, with total area around 0.2 $km^2$. The vegetation of the maya is dominated by *Echinochloa sp* (Difera) and *Ipomoea aquatica* (Hakim et al., 1979).

It was found during June 2016 survey that the maya remained with relatively shallow water depth and was covered with aquatic plants mostly *Ipomoea acquatica* (68%) and *Vossia cuspidata* (27%). *Ipomoea aquatica* is a member of the morning-glory family, and a fast-growing herbaceous that is commonly found floating in freshwater wetlands and water ponds (Patnaik, 1976). Inside the DNP, this plant is common in mayas with shallow water depths (Hassan, 2017). *Vossia cuspidata* is an aquatic spongy plant that inhabits the open water (Shaltout et al., 2005). Its stems may be submerged or floating. This aquatic plant forms dense floating mats of intertwined stems over water surfaces, covering and competing with the native submersed plants (Langeland and Burks, 1999). The tangled plant composition may negatively affect the maya by displacing native plants that are vital for fish and wildlife. The intertwined plants create dense impermeable covers over the maya's surface creating stagnant water environments that are ideal for breeding of mosquitoes.

***Gererrisa maya***

This maya is located at the western bank of the Dinder river at Long 12° 36' 28.5" N and Lat 34° 59' 22.7" E. It is a large (about 3.0 $km^2$) round shaped maya with a small island located at its centre. At the end of the 1970s, drought affected Gererrisa and it became dry. During 2001, the Wildlife authority of the park pumped water from a borehole to its centre, which helped the plants with water–loving plants such as *Cyperus* sp and *Nymphaea* to flourish again in the centre of the maya. It was noticed that *Cassia obtusifolia* had invaded it from the eastern side and along the canal from the borehole. Similar to *Ipomoea acquatica, Nymphaea* is also one of the common plants in mayas with shallow water (Hassan, 2017). During the 2016 survey, it was found that shallow water was available and *Ipomoea acquatica* was the most frequent grass (56%). Although *Cassia obtusifolia* was observed to cover parts of the maya, it was not recorded within the transects.

In summary, and in terms of water availability, the four surveyed mayas varied from full of water (Ras Amir) to shallow water (Abdelghani and Gererrisa) to dry condition (Musa). In terms of vegetation, all four mays covered by palatable green plant species (e.g. *Vossia cuspidate, Ipomoea aquatica, Boerhavia coccinea, Zornia glochidiata, Cyperus sp, Digera muricate* and *Cassia obtusifolia*). However, the report by Hakim et al. (1979) has shown that during the 1970s, the mayas were dominated by *Echinochloa spp*. The new invader "*Cassia obtusifolia*" is advancing in the periphery of some mayas, such as Ras

Amir. Although it is observed to cover parts of Gererrisa maya, but not recorded within the transects. This indicate that some of the native plants have been displaced by invasive plants that might not be suitable for wildlife and fish communities.

### 6.3.2 Delineation of water and vegetation surfaces results analysis

Remote sensing data were analyzed to identify the surface water extent as a result of flood for the purpose of wetland inundation detection and the related biomass using GIS tools and Landsat images. Musa maya was selected as a pilot maya.

Figure 6.1 shows that the results of the NDWI indicate no inundation during 2002. This is associated with very low flow of the Dinder river in 2002 (0.54 x $10^9$ $m^3/a$) as shown in Figure 6.3. In contrast, the year 2012 shows large inundation with total area reached 2.80 $km^2$. This is due to the high river flood (5.59 x $10^9$ $m^3/a$) and the construction of the connection canal in 2012. In 2013, due to low flood level, the inundated area declined to 1.27 $km^2$. Although the annual flow in 2015 was below average (0.95 x $10^9$ $m^3/a$), the Musa maya still inundated with a water extent of 0.85 $km^2$. This is likely due to the construction of the connection canal in 2012, which enhanced the filling of the maya. These results indicate that maya inundation is strongly linked to the flood magnitude of the Dinder river and its tributaries.

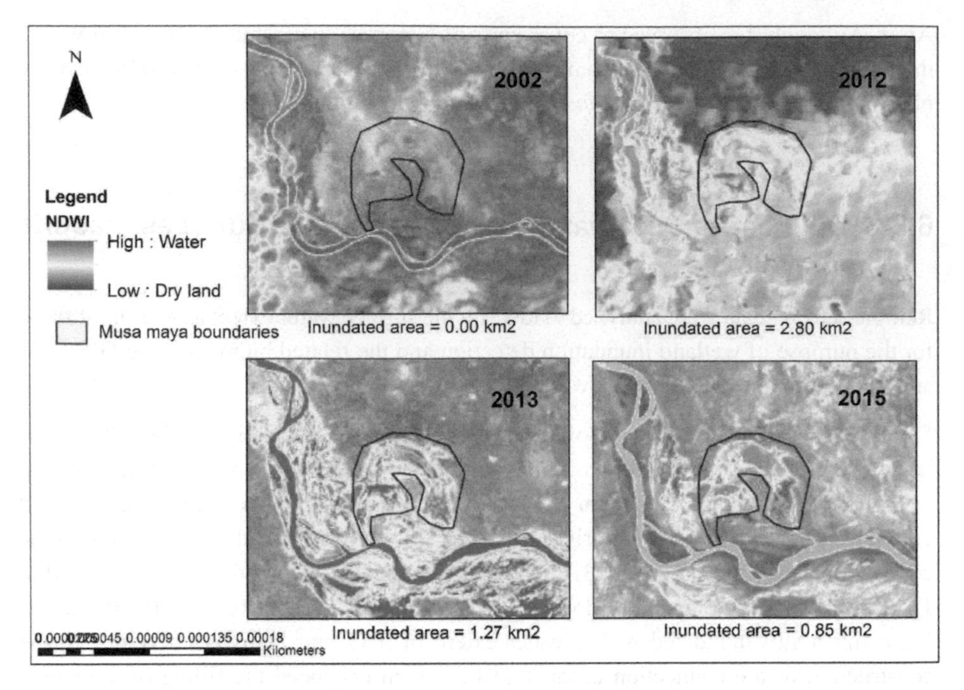


*Figure 6.1: Estimation of total inundated area of Musa maya using NDWI index.*

Figure 6.2, which presents the results of the NDVI, shows that the vegetation coverage in Musa maya during 2003 is about 1.00 km$^2$. This is associated with the very low flow of the Dinder river in 2002 (0.54 x 10$^9$ m$^3$/a) which led to a no inundation situation. In contrast, 2013 shows large vegetation coverage with total area reached 2.50 km$^2$. This was due to the large flood magnitude in 2012 (5.59 x 10$^9$ m$^3$/a) as shown in Figure 6.3. In 2014 the vegetation coverage declined to 1.56 km$^2$, due to a decline in the inundated area in 2013. Likewise, in 2016 the vegetation coverage declined to 1.31 km$^2$ following the decline in the inundation area in the flood season of 2015. These results indicate that maya vegetation coverage is strongly linked to the inundation extent.

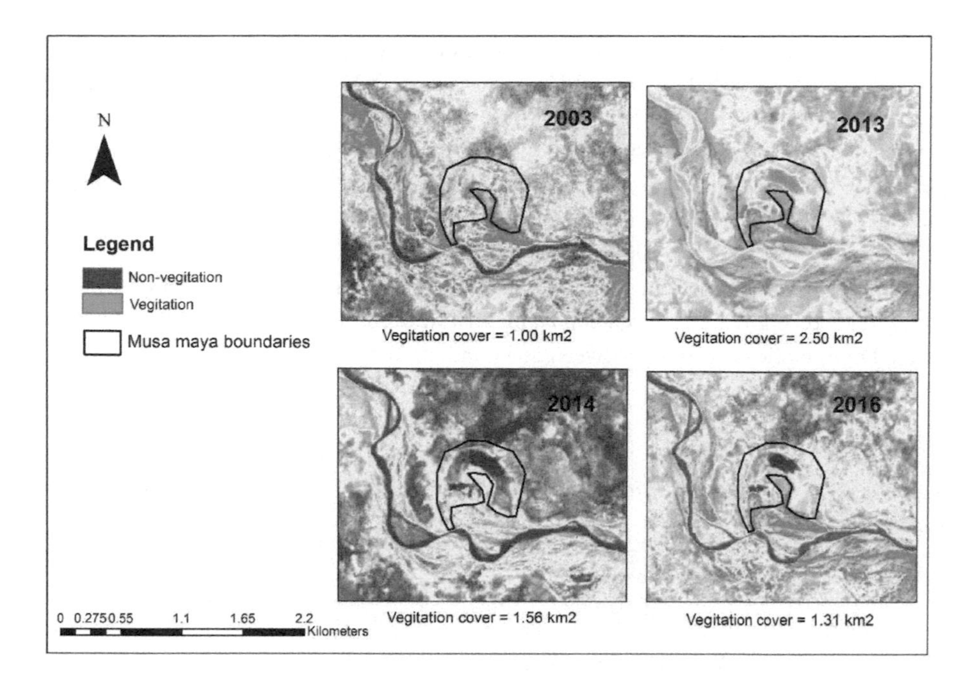


*Figure 6.2: Estimation of total vegetation coverage area of Musa maya using NDVI index.*

To estimate the probability of occurrence for a particular hydrological year in terms of average, wet and dry, the hydrological data for 117 years (please refer to Figure 2.5 in section 2.2.1) are classified into five categories as shown in Table 6.3. The approach placed the midpoints of the wet, average, and dry conditions at $30^{th}$, $50^{th}$, and $70^{th}$ percentiles, respectively. The high flow zone was centered at the $10^{th}$ percentile, while the low flow zone was centered at the $90^{th}$ percentile.

For 10% of the time, the annual streamflow equaled or exceeded $4.27 \times 10^9$ $m^3/a$, at 50% it equaled or exceeded $2.0 \times 10^9$ $m^3/a$ and at 90% flow equaled or exceeded $0.31 \times 10^9$ $m^3/a$.

The annual streamflow (1900-2016) recorded 12 years (10% exceedance) of high flow volumes; 47 years (50% exceedance) of average flow volumes; and 11 years (90% exceedance) of low flow volumes (Table 6.3). The annual average flow was between $2.0 \times 10^9$ $m^3/a$ and $3.0 \times 10^9$ $m^3/a$.

Based on these hydrological classifications and from the results of our previous study on hydrodynamics and morphology of the mayas, we found that Musa maya is inundated only if the Dinder river flow condition is above the average flow (30% exceedance) which is equaled or exceeded 3.19 x $10^9$ $m^3/a$. Out of 82 years (1935-2016) since the establishment of the DNP in 1935, the Musa maya is inundated in 22 years and remains dry in 60 years. The longest dry condition of the maya is found to be in twelve consecutive years (1976-1987) which is likely due to the devastating drought in the region during this period of time. This was followed by other dryness conditions in consecutive years during two periods 1989-1998 and 2000-2006. Figure 6.3 shows the conditions of the Musa maya in terms of wet and dry from 1935 to 2016. Inundation and dryness conditions in Musa maya in the recent years between 2000 and 2010 were confirmed by the wildlife personnel in the DNP, while conditions between 2011 and 2016 were confirmed through our direct observation during our yearly visits to the DNP.

*Table 6.3: Hydrologic conditions and probability of occurrence of Dinder river flow during the period 1900-2016*

| Category | Hydrologic-year type | Number of Years |
|---|---|---|
| *1* | Wet (10% exceedance) | 12 |
| *2* | Above Average (30% exceedance) | 24 |
| *3* | Average (50% exceedance) | 47 |
| *4* | Below Average (70% exceedance) | 23 |
| *5* | Dry (90% exceedance) | 11 |
| | Total | 117 |

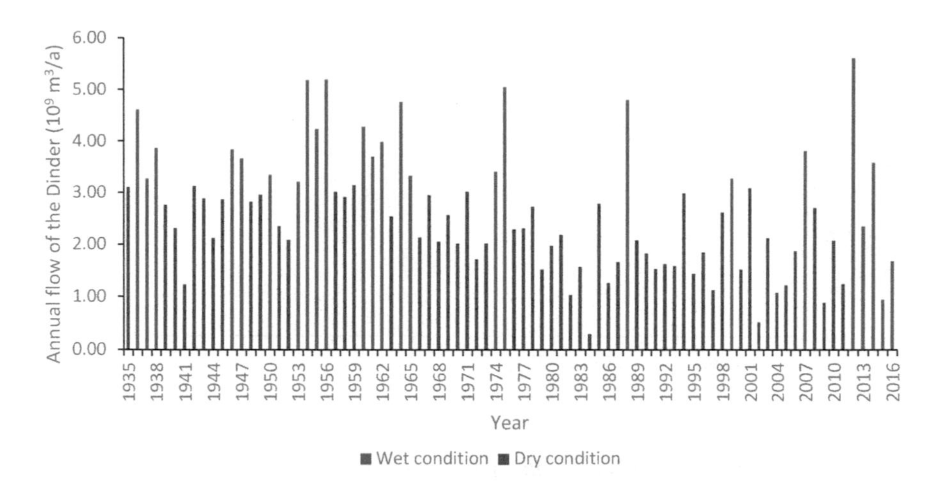


*Figure 6.3: The conditions of the Musa maya in terms of wet and dry during the dry seasons (January-June) since the establishment of the DNP in 1935.*

## 6.3.3 Wildlife census

***Wildlife census of 2010***

On the ground and through field work, the river ecosystem and the mayas ecosystem proved to have greater animal diversity than the Dahara (*Acacia seyal-Balanites aegyptiaca*) ecosystem. The first two ecosystems are characterized by the availability of three elements of high importance for wildlife namely; water, pasture and shade. This explains the presence of the majority of wildlife in these two ecosystems.

The census results include twenty-one species of mammals and one large bird that is the Ostrich. In the maya ecosystems, data for wildlife in four of the main mayas were analyzed. Figure 6.4 shows that Ras Amir maya has the greater wildlife count. In contrast, the Musa maya has the smaller count. This is likely due water availability in Ras Amir maya and the dry condition in Musa maya in 2010. Figure 6.5 illustrates the conditions of the two mayas during the dry period of 2011.

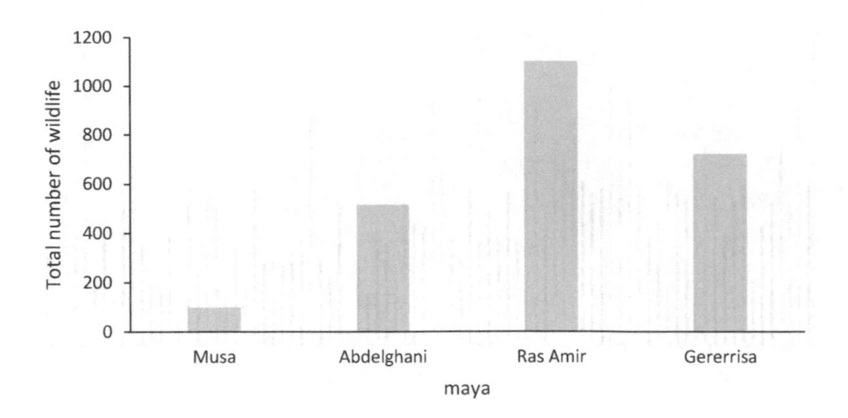


*Figure 6.4: Results of the census for the wildlife in four main mayas in the DNP from 11 to 12 May 2010. Source (Mubarak, 2010)*

*Figure 6.5: The conditions of a) Ras Amir maya and b) Musa maya during the dry season of 2011. Ras Amir was completely full of water, while Musa maya was dry (Pictures taken by Khalid Hassaballah, February 2011).*

***Wildlife census of 2016***

The population number and structure of Warthog in the four surveyed mayas are shown in Table 6.4. The results show that Gererrisa maya is highly populated with Warthog due to availability of water and pasture. Figure 6.6 shows the condition of Grerrisa maya during the dry period in April 2016. Abdelghani maya has a smaller population due to the little amount of available water and pasture observed during the census. We note that this census comes after the dry hydrological year 2015 with total annual flow of about 0.95 x $10^9$ $m^3/a$ compared to the long-term annual flow of 2.70 x $10^9$ $m^3/a$. Despite this, Musa maya recorded the second-highest population number after Gererrisa maya. This is likely due to the construction of the connection canal in June 2012 that enhanced the diversion of water from the Dinder river to the Musa maya, and hence increased the pasture area. But due to the low magnitude of river flow during the flood season of 2015, the Musa maya received very little water and thus dried up soon after the flood season. The maya was observed completely dry during the first week of April 2016. On the other hand, little water was found in Ras Amir maya during the same period (Figure 6.7).

Looking at the population structure in the four mayas, we noticed that the ratio of young (38%) to female (32%) is high. This indicates that production of young during this season is high and some females produce more than one young or some of the females were hunted by predators such as lions.

*Table 6.4: Population number and structure of Warthog. Source: Hassan (2017)*

| *Name of transect* | *Male* | *Female* | *Young* | *Total* |
|---|---|---|---|---|
| *Gelagu – Gererrisa* | 15 | 16 | 20 | 51 |
| *Gelagu –Musa* | 13 | 14 | 17 | 44 |
| *Gelagu – Ras Amir* | 10 | 12 | 11 | 33 |
| *Gelagu – Abdelghani* | 8 | 8 | 11 | 27 |
| *Total* | 46 | 50 | 59 | 155 |
| *Percentages* | 30% | 32% | 38% | 100% |

*Figure 6.6: Condition of Grerrisa maya during the dry period. The maya was completely full of water (Pictures taken by Khalid Hassaballah, April 2016).*

*Figure 6.7: The conditions of a) Ras Amir maya and b) Musa maya during the dry period of 2016. Little amount of water was found in Ras Amir maya. Musa maya was completely dry (Pictures taken by Khalid Hassaballah, April 2016).*

During the dry period of 2016, a wide range of burnt areas was observed in the DNP. Fires inside the DNP during the dry season and other human activities greatly affect the ecology of the area. For instance, both poachers and honey gatherers light fires throughout the DNP (Hassaballah et al., 2016). Many of the fires originated and are certainly set outside the park by cultivators, honey collectors and nomad pastoralists seeking to reduce the grass cover in order to improve access of livestock to perennial grasses (Yousif and Mohamed, 2012). The DNP personnel also set fires when opening roads at the beginning of the dry season. It is generally admitted by the DNP personnel and through our observations in the DNP between 2010-2016 that most of the park burns nearly every year and particularly during consecutive years of drought (Figure 6.8). The park staff can do little to control these fires without firefighting equipment.

*Figure 6.8: Wide range of burnt areas observed in the DNP. (Pictures taken by Khalid Hassaballah, April 2016).*

To determine the distribution and preference to different habitat for wildlife in the DNP in terms of burnt (dry lands) and unburnt area (wetlands), data on Warthogs from the census conducted by Hassan (2017) were used as an indicator. The distribution and preference of Warthog to different habitats is shown in Table 6.5. 84% of the total populations were found within the wetlands area (mayas) compared to only 16% within the burnt area. This indicates that Warthogs prefer grassland around the mayas rather than burnt and dry areas. This is likely due to availability of water, pasture and shelter and to hide from predators.

*Table 6.5: Distribution and preference of Warthogs to different habitats in the DNP during February-May 2016. Source: Hassan (2017)*

| Name of transect | Grassland | Woodland | Riverine Forest | Burnt dry areas | Maya wetlands | Total |
|---|---|---|---|---|---|---|
| Gelagu – Gererrisa | 28 | 23 | - | 9 | 42 | 51 |
| Gelagu – Musa | 28 | 19 | - | - | 44 | 44 |
| Gelagu – Ras Amir | 20 | 13 | - | 9 | 24 | 33 |
| Gelagu – Abdelghani | 17 | 10 | 2 | 7 | 20 | 27 |
| Total | 93 | 65 | 2 | 25 | 130 | 155 |
| Percentages of burned to unburned area (%) | | | | 16 | 84 | 100 |

## 6.4 Conclusions

Maya wetlands are important hydrological resources that support communities of plants and wildlife in Dinder National Park. To assess the condition of functioning of the mayas, a systematic-random quadrat (SRQ) method was used to collect flora's data (indicators) from four mayas inside the DNP. In addition, the NDWI was used to estimate the inundation extent and the NDVI was used to estimate the related vegetation coverage in the pilot Musa maya. Data on wildlife censuses in the four mayas were analyzed and its relations to hydrological variability and vegetation cover were identified.

The SRQ survey distinguished seven plant species in the four surveyed mayas, with floristic composition of plant species that considerably varies across the studied mayas.

The aquatic plants noted in this study have certain features in common, such as vegetative reproduction and relatively rapid growth. It has also been observed that most of the plant communities in terrestrial and aquatic habitats are often overwhelmingly dominated by one species. As an example, *Ipomoea acquatica* dominates in the shallow open water of Abdelghani maya and *Cassia obtusifolia* dominates in the deep-water edge of Ras Amir maya. The occurrence of plant communities dominated by single species results in reduced coverage of the less competitive species, and hence a decline in the species diversity of that particular community (Mohler and Liebman, 1987). The same results have been derived by Shaltout et al. (1995) and Shaltout et al. (2005) in their studies on the vegetation of the Mediterranean area of the Nile Delta and that of the vegetation-environment relationships in south Nile Delta.

The NDWI results indicate that maya inundation patterns are strongly linked to the flood magnitude of the Dinder river and its tributaries. Similarly, the NDVI show that the vegetation coverage in the maya is strongly linked to the inundation extent. However, other factors that affected the vegetation coverage in the park such as incursion of livestock due to the seasonal movements of pastoralists and uncontrolled deliberate and non-deliberate fires observed during the survey cannot be excluded.

The wildlife censuses have shown that the population size and distribution of wildlife in the DNP are related to the availability of water and pasture which are affected by the hydrological variability. 84% of the total wildlife (herbivores) observed were found in the grassland within the periphery of mayas compared to only 16% in the burnt and open areas. This is likely due to availability of water, pasture and shelter.

The findings of this study are relevant for water management, wildlife conservation and the rehabilitation and restoration activities currently being implemented in the DNP.

Only four mayas were surveyed in this study. Studies of more mayas will give a more comprehensive overview of flora and fauna distribution in the DNP. At the same time periodic monitoring and assessment of flora and fauna is important for quantifying changes in the status of the ecology of mayas.

# 7

# CONCLUSIONS AND RECOMMENDATIONS

## 7.1 Conclusions

Assessment of spatial and temporal variability of the hydro-climatic variables and links to land use and land cover change is key to understand the eco-hydrological interactions in the Dinder and Rahad basins, the host of the precious ecosystem of the Dinder National Park (DNP). However, this is a typically data scarce region with extremely limited data on climate, hydrology and the ecosystem. LULC changes are posing extra challenges for comprehensive assessment. The hydrological processes of the basin are not fully understood under recent and current conditions, furthermore, prediction of hydrological and morphological dynamics in the future remain challenging and uncertain. Therefore, further in-depth hydrological and morphological studies of the basins and their interactions with the ecosystem are essential to inform better management of water resources and ecosystems in the Dinder and Rahad basins.

This dissertation investigates the impacts of land degradation on the hydrology and morphology of the Dinder and Rahad rivers, and interlinkages to the ecohydrology of the DNP, Sudan. The study followed a structured approach that includes: long-term trend analysis of the hydroclimatic data to detect whether the hydroclimatic conditions have changed, and hence impacted the hydrology of the Dinder and Rahad basins, or if observed changes can be attributed to large scale LULC changes in the basins. The latter has been investigated by detecting changes from satellite imageries during the last four decades. Then, the study attempted to assess the filling/emptying of the maya wetlands in the DNP. Finally, the study assessed the ecohydrological interlinkages in maya wetlands and how they are influenced by the hydrology and morphology of the Dinder and Rahad rivers. The study uses an ensemble of techniques and tools, including: statistical analysis, land use and land cover change detection analysis, field measurements, rainfall-runoff modelling, GIS and remote sensing data acquisition and analysis, hydrodynamic and morphological modelling, and desktop studies to synthesize the understanding of the ecohydrology of the maya wetlands .

It was the first time the ecohydrology of the Dinder and Rahad basins has been studied.

The following conclusions have been written to directly address the four research questions of the dissertation (see chapter 1).

### 7.1.1 Is there any significant long-term trend in the hydroclimatic variables of both Dinder and Rahad rivers, and if so to what extent?

The long-term trends in hydro-climatology of the Dinder and Rahad sub-basins, of the Blue Nile, Ethiopia/Sudan were assessed. The non-parametric Mann-Kendall (MK) and Pettitt tests were applied to analyze the trends and the change points of time series of streamflow, rainfall and temperature. Trends have been assessed at 5% significance level for different time periods and varying lengths based on data availability. The long-term trend of the Dinder and Rahad hydro-climatology has been analyzed for: rainfall (12 stations), temperature (2 stations), and streamflow (2 stations), over different periods of time. The mean annual temperature showed statistically significant increasing trends at the rate of about 0.24 and 0.30 °C/decade in Gedarif and Gonder stations, respectively. No significant change in rainfall has been detected. The results of the trend analysis of rainfall agree with the literature for the neighboring catchment of the Blue Nile (e.g. Tesemma et al., 2010; Gebremicael et al., 2013; Tekleab et al., 2013).

The mean annual streamflow of the Rahad river exhibited a statistically significant increasing trend, but not for the Dinder river which showed no significant changes. The trend of the monthly mean flows showed significant increasing trends in Rahad river for July, August and November, while no significant trend was observed in Dinder river. The monthly maxima flow showed a significantly decreasing trend of August maxima flows and decreasing trend of November maxima flows in the Dinder river, while no evidence for a significant trend of monthly maxima flows of the Rahad river. Reduction of the Dinder peak flow can have a direct impact on filling of the mayas, being the main water source for the DNP during the dry months. The Pettitt test indicates that the changing points of streamflow in the Dinder and Rahad occurred during the late 1980s and the early 1990s. The increasing temperature associated with increasing flow in Rahad river indicates that the increasing trend of temperature may not necessarily imply a decreasing runoff since land use and land cover is another factor in controlling the partitioning of rainfall.

The IHA-based analysis has shown that the flow of the Rahad river was associated with significant upward alterations in some of the hydrological indicators, while the flow of the Dinder river was associated with significant downward alterations. Particularly, these were: a) a decrease in the magnitude of the Dinder river flow during August (peak flow) and an increase in low flows (November); b) a decrease in magnitude of the Dinder flow extremes (i.e. 1, 7, 30 and 90-day maxima); and c) a decrease in the Dinder flow rise rate and an increase in flow fall rate.

The result of the trend analysis showed no significant long-term changes of rainfall over Dinder and Rahad basins. However, temperature showed a significant increase, while the runoff showed increasing/decreasing trends. This means that other factors than climate variability (e.g. land use and land cover changes) might be responsible for streamflow alterations. This also means that our hypothesis related to the impacts of the long-term trends in climatic variables on hydrological changes, that was stated in section 1.5 is proven to be not true.

### 7.1.2 What are the impacts of the land use and land cover changes in the upper Dinder and Rahad on the catchment runoff response?

After analyzing and demonstrating the limited impact of climate factors as drivers for hydrological change in the D&R basins, the study continued to assess whether LULC change influences the hydrology of the basin. Analysis of streamflow response to land use and land cover changes using satellite data and hydrological modelling was performed. First, LULC changes within the D&R basins were assessed using satellite images for the years 1972, 1986, 1998 and 2011. The accuracy assessment of the supervised land cover classification shows reliable classification results. The overall LULC classification accuracy levels for the four images ranged from 82% to 87%, with Kappa indices of agreement ranging from 77% to 83%. In general, the results showed relatively large decrease in woodland and large increase in cropland. LULC changes between 1972 and 2011, the woodland decreased from 42% to 14 % in Dinder and from 35% to 14 % in Rahad. The cropland increased from 14% to 47 % in Dinder and from 18% to 68 % in Rahad, respectively. The rate of deforestation was high during the period 1972–1986, and can likely be attributed to a number of factors: the severe drought of 1984–1985; expansion of agricultural land; increased demand for wood as fuel and for construction due to the increase in population. On the other hand, the increase in woodland from 23% to 27% in Dinder and from 14% to 21% in Rahad during the period between 1986 and 1998 is probably due to reforestation activities in the basin or due to natural environmental recovery. Nevertheless, the LULC change detection analysis have shown that the magnitude of deforestation is still much larger than the reforestation. The cropland expansion over the period 1986–1998 (from 15% to 45% in Dinder and from 26% to 55% in Rahad) was larger than the cropland expansion over the period 1998–2011 (from 45% to 47% in Dinder and from 55% to 68% in Rahad), suggesting that most of the areas that are suitable for cultivation have most likely been occupied, or land tenure regulations have controlled the expansion of cultivation by local communities.

A hydrological model based on a Wflow hydrologic model (Schellekens, 2011), has been applied for different LULC scenarios of the D&R basins. The model results showed that the LULC changes were significantly increased the streamflow during the years 1986 and 2011, particularly in the Rahad river. This could be attributed to the severe land degradation during 1984–1985 and the large expansion in cropland in the Rahad catchment to 68% of the total area in 2011. The IHA-based analysis verified the effects of LULC changes on flow alterations, and showed significant upward/downward alterations of the Dinder and Rahad flows in terms of magnitude, timing and rate of change of river flows. This proved that our hypothesis related to the impacts of land use and land cover changes on the hydrology of Dinder and Rahad basins that was stated earlier (section 1.5) is true.

The alterations in the streamflow characteristics (i.e. magnitude, duration, timing, and rate of change) have direct effects on the flow regime of the Dinder and Rahad rivers, and hence on the species depending on river flow dynamics. In other words, alterations of the annual floods of the Dinder river, and hence reduction of river spills into the mayas reduces the production of native river–floodplain fauna and flora and the migration of animals that may be connected to mayas inundation.

### 7.1.3 How does filling and emptying of mayas normally occur, and what are the key factors controlling the processes?

After analyzing whether LULC change influences the hydrology of the basin or not, the study continued to assess the effect of the hydrodynamic and morphological processes on the maya wetlands in terms of filling/emptying and sediment transport processes.

To understand the process of filling/emptying of the maya wetlands, a quasi 3D morphodynamic model was built for a pilot area called Musa maya. The model investigates the effect of morphological changes on the Dinder river and the maya. The model extent covered an area of about 105 km$^2$ inside the DNP. SRTM (90 m) was used along a 20-km reach of the Dinder river.

A network of two Divers for water level measurement was established in June 2013 to collect data for this study, as there was not any hydrological data at site. The divers were set to take measurements automatically on hourly basis at the river and at the Musa maya. The observed water levels were used to calibrate the hydrodynamic model.

To understand the hydrological and morphological connectivity of the maya in terms of filling/emptying and sediment transport processes, six scenarios were investigated by the morphodynamic model. The first three scenarios consider three different hydrologic conditions of wet, average, and dry years for the existing system with the constructed connection canal. While, the other three scenarios assume the same hydrology but for the natural system without the connection canal.

It has to be mentioned that the approximation of the topography (based on SRTM data corrected with land surveys), creates significant uncertainty of the bed topography. This is reflected by the morphological predictions. Based on model results for the six scenarios, the following major conclusions could be drawn:

i. The hydrodynamics and sedimentology of the Musa maya are driven by the following two factors: a) the hydrological variability of the Dinder river; and b) sediment deposits on the bed channel of the natural drainage (feeder) that is likely caused by settled dead trees.
ii. Three phases of hydrological connectivity were distinguished: 1) filling phase, 2) drainage phase, and 3) isolation phase.
iii. The modelling results have shown that the constructed connection canal has enhanced the filling mechanism of the maya, particularly in the dry years. However, the field measurement has shown a sever bank erosion on the right bank of the river cross-section just downstream of the connection canal (about 60 m in three years).
iv. Modification of river width create more room for water within the river reach, and may lead to reduction in the water flows to the maya.

These results proved that our hypothesis related to the effects of the hydrological alterations and morphological changes on the filling mechanism of the maya wetlands of the DNP, that was stated earlier (section 1.5) is true.

This study also highlighted the value of integrating field observations and morphodynamic modelling for understanding the flooding and sediment transport processes in the maya wetlands. However, considering the high uncertainty in the data and model, the hydrological and morphological quantification should be considered as indicators for the effects of possible future hydrological and morphological changes, rather than exact values of the expected changes.

The methodology applied in this study can be applied to other maya wetlands with similar hydrological conditions as the Musa maya, in particular, the mayas that are located along the Dinder river. However, it requires more reliable measurements of topography and hydrology before a robust 3D model can be built.

### 7.1.4 Can identified changes of the maya wetlands functioning (i.e., filling and emptying) be related to the local ecosystem (e.g. flora and fauna)?

The ecohydrology of the maya wetlands in the DNP is assessed and relations between vegetation dynamics, wildlife and water availability were identified. Field data on vegetation composition and wildlife were collected from four mayas to assess the ecosystem status and patterns of changes. To determine the status of functioning of the mayas, a systematic-random quadrat (SRQ) method was used to collect flora's data (indicators) from four mayas inside the DNP. In addition, the normalized difference water index (NDWI) was used to estimate the inundation extent and the normalized difference vegetation index (NDVI) was used to estimate the related vegetation coverage in the pilot Musa maya. Data on wildlife censuses in the four mayas were analyzed and its relations to hydrological variability and vegetation cover were identified.

The SRQ survey distinguished seven plant species in the four surveyed mayas, with floristic composition of plant species that considerably varies across the studied mayas.

The aquatic plants noted in this study have certain features in common, such as vegetative reproduction and relatively rapid growth. It has also been observed that most of the plant communities in terrestrial and aquatic habitats are often overwhelmingly dominated by one species. As an example, *Ipomoea acquatica* dominates in the shallow open water of Abdelghani maya and *Cassia obtusifolia* dominates in the deep-water edge of Ras Amir maya. The occurrence of plant communities dominated by single species results in reduced coverage of the less competitive species, and hence a decline in the species diversity of that particular community (Mohler and Liebman, 1987). The same results have been derived by Shaltout et al. (1995) and Shaltout et al. (2005) in their studies on the vegetation of the Mediterranean area of the Nile Delta and that of the vegetation-environment relationships in south Nile Delta.

The NDWI results indicate that maya inundation patterns are strongly linked to the flood magnitude of the Dinder river and its tributaries. Similarly, the NDVI show that the vegetation coverage in the maya is strongly linked to the inundation extent. However, other factors that affected the vegetation coverage in the park such as incursion of livestock due to the seasonal movements of pastoralists and uncontrolled deliberate and non-deliberate fires observed during the survey cannot be excluded.

The wildlife censuses have shown that the population size and distribution of wildlife in the DNP are related to the availability of water and pasture which are affected by the hydrological variability. 84% of the total wildlife (herbivores) observed were found in the grassland within the periphery of mayas compared to only 16% in the burnt and open areas. This is likely due to availability of water, pasture and shelter.

The findings of this study are relevant for water management, wildlife conservation and the rehabilitation and restoration activities currently being implemented in the DNP. The findings are also proved that our hypothesis related to the effects of water availability on vegetation dynamics and wildlife in mayas ecosystem of the DNP, that was stated earlier (section 1.5) is true.

Only four mayas were surveyed in this study. Studies of more mayas will give a more comprehensive overview of flora and fauna distribution in the DNP. At the same time periodic monitoring and assessment of flora and fauna is important for quantifying changes in the status of the ecology of mayas.

In general, this dissertation provided a scientifically important and practically relevant example of hydrological and morphological connectivity assessment, their linkage to ecohydrology and their use in water resources planning and management in a transboundary river basin context, which is useful for the Dinder and Rahad and other transboundary river basins in the region and worldwide. However, the study has two major limitations:

i. The study uses complex models in data scarce catchments. As such, calibration and validation performance of the models, particularly that of the hydrodynamic and morphology, was not very high. Detailed hydrodynamic and morphology modeling and analysis of the filling/emptying of the maya wetlands will require further fine-tuning of the models based on more complete data sets.
ii. The use of integrated models, besides the lack of reliable and sufficient data, introduces model uncertainties. Unfortunately, such unavoidably uncertainties are difficult to quantify in integrated modeling in general, and were not quantified in this study.

## 7.2 Recommendations

This study analyzed the ecohydrology of the maya wetlands of the DNP and how they are affected by the hydrology and morphology of the Dinder and Rahad rivers. Detailed insights have been obtained on the drivers for change, and their implications on the mayas ecosystem. Some of the results are directly relevant for water resources planning in the region and for the conservation of mayas in the DNP, while other results need further consolidation and investigations including detailed field surveys and modelling studies.

Strengthening of hydro-climatic monitoring networks is recommended to improve data availability for hydrological, hydrodynamics, morphological and other water management models to support decision-making processes. In this regard, the study recommends installation of continuous hydrological monitoring stations in the DNP. This should include water level, flow and sediment measurements during the flood season. Given the inaccessibility during the rainy season, automatic recorders are necessary in the context of the DNP.

The results of the land use and land cover changes indicate that LULC changes play a more considerable role for changes in the streamflow hydrograph than the climate variability. As part of the Dinder and Rahad basins are located in Ethiopia, regional and transboundary basin approaches are necessary. Therefore, intensive soil conservation measures upstream aiming at reducing land degradation (e.g. implementing soil and water conservation structures and adopting appropriate farming practices) are recommended for sustainable land and water management in the Dinder and Rahad basins.

The study also recommends regular ecological survey of more mayas than the four mayas included in this study. The survey should be conducted by the Wildlife Research Center and should cover plant species, mammals, reptiles and birds. This is very important for quantifying the changes in the conditions of the mayas' ecology.

In general, the data scarcity problem in combination with often poor data quality should be mitigated using new measurement techniques to improved spatial and temporal coverage. Moreover, continuous data quality monitoring (e.g. regularly updating rating curves at the discharge gauges during different flow regimes, building capacities of human resources, who operate the hydrometeorological stations) would be essential to reduce the data uncertainty and enhance further hydrological research and better water management in the D&R basins.

In this regard, the Ethiopian Ministry of Water and Energy-Ethiopia, the Ministry of Irrigation and Water Resources-Sudan, the National Meteorological Agency-Ethiopia and the National Meteorological Authority-Sudan should work together on the data collection as well as measures related to quality assurance, archiving and agreement protocols for data sharing to be used for research in order to achieve mutual benefits.

Finally, effective management of Dinder and Rahad basins requires transboundary cooperation. Further studies should extend beyond the political border to include Alatish National Park on the other side of the border Sudan-Ethiopia. The Alatish is a trans-national park sharing boundaries with the Dinder National Park. The Alatish National Park (ANP) and the Dinder National Park share the largest tributaries of the Blue Nile, which are the Alatish, Ayima (Dinder) Gelagu and Shinfa (Rahad) rivers. As both parks are sharing common water resources, environment and human disturbance on ANP's water resources (e.g. illegal fishers who poison the water bodies by using organic and synthetic compounds to harvest fish easily) will of course affect the aquatic community of the DNP as the water flows down from ANP to the DNP.

# REFERENCES

Abdel Hameed, S. M. (1998). BIOSPHERE RESERVES IN THE SUDAN. Nature et Faune Wildlife and Nature 14: 18-31.

Abdel Hameed, S. M., A. A. Hamid, N. M. Awad, E. E. Maghraby, O. O.A. and H. S.H. (1996a). Assessment of Wildlife Habitats in Dinder National Park by Remote Sensing Techniques. Albuhuth Vol. 5(1): pp41-55.

AbdelHameed, S. M., N. M. Awad, A. I. ElMoghraby, A. A. Hamid, S. H. Hamid and O. A. Osman (1997). Watershed management in the Dinder National Park, Sudan. Agricultural and Forest Meteorology 84(1): 89-96.

Al Fugara, A. M., B. Pradhan and T. A. Mohamed (2009). Improvement of land-use classification using object-oriented and fuzzy logic approach. Applied Geomatics 1(4): 111-120.

Alper, J. (1998). Ecosystem'engineers' shape habitats for other species. Science 280(5367): 1195-1196.

Alvarez-Mieles, G., K. Irvine, A. Griensven, M. Arias-Hidalgo, A. Torres and A. E. Mynett (2013). Relationships between aquatic biotic communities and water quality in a tropical river–wetland system (Ecuador). Environmental Science & Policy 34: 115-127.

Armah-Agyeman, G., J. Loiland, R. Karow and B. Bean (2002). Dryland cropping systems: Sudangrass, Corvallis, Or.: Extension Service, Oregon State University. EM 8793.

Arthington, Á. H., R. J. Naiman, M. E. Mcclain and C. Nilsson (2010). Preserving the biodiversity and ecological services of rivers: new challenges and research opportunities. Freshwater biology 55(1): 1-16.

Baird, A. J., J. S. Price, N. T. Roulet and A. L. Heathwaite (2004). Special issue of hydrological processes wetland hydrology and eco-hydrology. Hydrological Processes 18(2): 211-212.

Balica, S., I. Popescu, L. Beevers and N. G. Wright (2013). Parametric and physically based modelling techniques for flood risk and vulnerability assessment: a comparison. Environmental modelling & software 41: 84-92.

Basheer, A. K., H. Lu, A. Omer, Abubaker B and A. M. S. Abdelgader (2016). Impacts of climate change under CMIP5 RCP scenarios on the streamflow in the Dinder River and ecosystem habitats in Dinder National Park, Sudan. Hydrology and Earth System Sciences 20(4): 1331-1353.

Bates, P. D. and A. P. J. De Roo (2000). A simple raster-based model for flood inundation simulation. Journal of Hydrology 236(1–2): 54-77.

Berhanu, K. and E. Teshome (2018). Opportunities and challenges for wildlife conservation: The case of Alatish National Park, Northwest Ethiopia. African Journal of Hospitality, Tourism and Leisure 7(1): 1-13.

Bernhardt, E. S. and M. A. Palmer (2011). River restoration: the fuzzy logic of repairing reaches to reverse catchment scale degradation. Ecological applications 21(6): 1926-1931.

Bewket, W. and G. Sterk (2005). Dynamics in land cover and its effect on stream flow in the Chemoga watershed, Blue Nile basin, Ethiopia. Hydrological processes 19(2): 445-458.

Bewket, W. and E. Teferi (2009). Assessment of soil erosion hazard and prioritization for treatment at the watershed level: Case study in the Chemoga watershed, Blue Nile basin, Ethiopia. Land Degradation & Development 20(6): 609-622.

Bhandari, A., A. Kumar and G. Singh (2012). Feature extraction using Normalized Difference Vegetation Index (NDVI): a case study of Jabalpur city. Procedia Technology 6: 612-621.

Biro, K., B. Pradhan, M. Buchroithner and F. Makeschin (2013). Land use/land cover change analysis and its impact on soil properties in the northern part of Gadarif region, Sudan. Land Degradation & Development 24(1): 90-102.

Block, P. and B. Rajagopalan (2006). Interannual variability and ensemble forecast of Upper Blue Nile Basin Kiremt season precipitation. Journal of Hydrometeorology 8: 327-343.

Block, P. J., K. Strzepek and B. Rajagopalan (2007). Integrated Management of the Blue Nile Basin in Ethiopia, IFPRI Discussion Paper.

Bond, B. (2003). Hydrology and ecology meet-and the meeting is good. Hydrological Processes 17(10): 2087-2089.

Boon, P. (1992). Essential elements in the case for river conservation. River conservation and management: 11-33.

Bracken, L. J. and J. Croke (2007). The concept of hydrological connectivity and its contribution to understanding runoff-dominated geomorphic systems. Hydrological Processes: An International Journal 21(13): 1749-1763.

Bracken, L. J., L. Turnbull, J. Wainwright and P. Bogaart (2015). Sediment connectivity: a framework for understanding sediment transfer at multiple scales. Earth Surface Processes and Landforms 40(2): 177-188.

Brierley, G., K. Fryirs and V. Jain (2006). Landscape connectivity: the geographic basis of geomorphic applications. Area 38(2): 165-174.

Brookes, A. and F. D. Shields (1996). River channel restoration: guiding principles for sustainable projects, J. Wiley.

Bruno, J. F. (2001). Habitat modification and facilitation in benthic marine communities. Marine community ecology.

Burn, D. H. and M. A. Hag Elnur (2002). Detection of hydrologic trends and variability. Journal of Hydrology 255(1–4): 107-122.

Cairns, J., P. V. McCormick and B. Niederlehner (1993). A proposed framework for developing indicators of ecosystem health. Hydrobiologia 263(1): 1-44.

Carignan, V. and M.-A. Villard (2002). Selecting indicator species to monitor ecological integrity: a review. Environmental monitoring and assessment 78(1): 45-61.

Cigizoglu, H., M. Bayazit and B. Önöz (2005). Trends in the maximum, mean, and low flows of Turkish rivers. Journal of Hydrometeorology 6(3): 280-290.

Congalton, R. G. and K. Green (2008). Assessing the accuracy of remotely sensed data: principles and practices, CRC press.

Connelly, N. A. and B. A. Knuth (2002). Using the coorientation model to compare community leaders' and local residents' views about Hudson river ecosystem restoration. Society &Natural Resources 15(10): 933-948.

Cook, B. J. and F. R. Hauer (2007). Effects of hydrologic connectivity on water chemistry, soils, and vegetation structure and function in an intermontane depressional wetland landscape. Wetlands 27(3): 719-738.

Crain, C. M. and M. D. Bertness (2006). Ecosystem engineering across environmental gradients: implications for conservation and management. BioScience 56(3): 211-218.

Dale, V. H. and S. C. Beyeler (2001). Challenges in the development and use of ecological indicators. Ecological indicators 1(1): 3-10.

Dasmann, W. (1972). Development and management of the Dinder N. Park and its wildlife: A report to the Government of Sudan. Rome, FAO No TA 31 1 3: 61p.

De Steven, D. and M. M. Toner (2004). Vegetation of upper coastal plain depression wetlands: environmental templates and wetland dynamics within a landscape framework. Wetlands 24(1): 23-42.

Deep, S. and A. Saklani (2014). Urban sprawl modeling using cellular automata. The Egyptian Journal of Remote Sensing and Space Science 17(2): 179-187.

DeFries, R. and L. Bounoua (2004). Consequences of land use change for ecosystem services: A future unlike the past. GeoJournal 61(4): 345-351.

DeFries, R. and K. N. Eshleman (2004). Land-use change and hydrologic processes: a major focus for the future. Hydrological processes 18(11): 2183-2186.

Deltares (2010). Delft 3D-Flow User manual, Simulation of Multidimensional Hydrodynamic Flows and Transport Phenomena, Delft, the Netherlands, 38–43,.

Deursen, W. P. A. (1995). Geographical information systems and dynamic models: development and application of a prototype spatial modelling language, Faculteit Ruimtelijke Wetenschappen, Universiteit Utrecht.

Dunlop, E. A. (2007). Mapping and modelling the invasion dynamics of Senna obtusifolia at different levels of scale in Australia, Queensland University of Technology.

Elagib, N. A. (2010). Trends in intra-and inter-annual temperature variabilities across Sudan. Ambio 39(5-6): 413-429.

Elagib, N. A. and M. G. Mansell (2000). Recent trends and anomalies in mean seasonal and annual temperatures over Sudan. Journal of Arid Environments 45(3): 263-288.

Ellis, E. A., K. A. Baerenklau, R. Marcos-Martínez and E. Chávez (2010). Land use/land cover change dynamics and drivers in a low-grade marginal coffee growing region of Veracruz, Mexico. Agroforestry Systems 80(1): 61-84.

Elshamy, M. E., I. A. Seierstad and A. Sorteberg (2009). Impacts of climate change on Blue Nile flows using bias-corrected GCM scenarios. Hydrology and Earth System Sciences 13(5): 551-565.

FAO (1984). Ethiopian Highlands Reclamation Study (EHRS). Final Report, Vols. 1–2, Rome.

Fenner, M. (1982). Aspects of the ecology of Acacia-Commiphora woodland near Kibwezi, Kenya. East Africa Natural History Society.

Fernandes, J. N., J. B. Leal and A. H. Cardoso (2018). Influence of floodplain and riparian vegetation in the conveyance and structure of turbulent flow in compound channels. E3S Web of Conferences, EDP Sciences.

Gash, J. (1979). An analytical model of rainfall interception by forests. Quarterly Journal of the Royal Meteorological Society 105(443): 43-55.

Gash, J. H., C. Lloyd and G. Lachaud (1995). Estimating sparse forest rainfall interception with an analytical model. Journal of Hydrology 170(1-4): 79-86.

Gebremicael, T. G., Y. A. Mohamed, G. D. Betrie, P. van der Zaag and E. Teferi (2013). Trend Analysis of Runoff and Sediment Fluxes in the Upper Blue Nile Basin: A Combined Analysis of Statistical Tests, Physically-based Models and Landuse Maps. Journal of Hydrology 482: 57-68.

Ghil, M. and R. Vautard (1991). Interdecadal oscillations and the warming trend in global temperature time series. Nature 350(6316): 324-327.

Giller, P. S. (2005). River restoration: seeking ecological standards. Editor's introduction. Journal of applied ecology 42(2): 201-207.

Glińska-Lewczuk, K. (2009). Water quality dynamics of oxbow lakes in young glacial landscape of NE Poland in relation to their hydrological connectivity. Ecological Engineering 35(1): 25-37.

Gomoiu, M. (1998). Notes on the Fauna and Flora of the Danube Delta. NEAR Summer School.

Gumindoga, W., T. H. M. Rientjes, A. T. Haile and T. Dube (2014). Predicting streamflow for land cover changes in the Upper Gilgel Abay River Basin, Ethiopia: A TOPMODEL based approach. Physics and Chemistry of the Earth, Parts A/B/C 76–78: 3-15.

Gumiri, S. and T. Iwakuma (2002). The dynamics of rotiferan communities in relation to environmental factors: comparison between two tropical oxbow lakes with different hydrological conditions. Internationale Vereinigung für theoretische und angewandte Limnologie: Verhandlungen 28(4): 1885-1889.

Gurnell, A., C. Hupp and S. Gregory (2000). Linking hydrology and ecology. Hydrological Processes 14(16-17): 2813-2815.

Hakim, S., B. Fadlalla, N. M. Awad and S. A. Abdelwahab (1979). Ecosystems of the vegetation of Dinder National Park. Unpublished report, Wildlife Research Center, Khartoum, Sudan.

Hannah, D. M., P. J. Wood and J. P. Sadler (2004). Ecohydrology and hydroecology: A 'new paradigm'? Hydrological Processes 18(17): 3439-3445.

Hansen, A., R. DeFries and W. Turner (2004). Land Use Change and Biodiversity: A Synthesis of Rates and Consequences during the Period of Satellite Imagery Land Change Science: Observing, Monitoring, and Understanding Trajectories of Change on the Earth's Surface Springer Verlag, New York, NY.: 277-299.

Hassaballah, K., Y. Mohamed, S. Uhlenbrook and K. Biro (2017). Analysis of streamflow response to land use and land cover changes using satellite data and hydrological modelling: case study of Dinder and Rahad tributaries of the Blue Nile (Ethiopia–Sudan). Hydrol. Earth Syst. Sci. 21(10): 5217-5242.

Hassaballah, K., Y. A. Mohamed and S. Uhlenbrook (2016). The Mayas wetlands of the Dinder and Rahad: tributaries of the Blue Nile Basin (Sudan). The Wetland Book: II: Distribution, Description and Conservation. C. M. Finlayson, G. R. Milton, R. C. Prentice and N. C. Davidson. Dordrecht, Springer Netherlands: 1-13.

Hassaballah, K., Y. A. Mohamed and S. Uhlenbrook (2019). The long-term trends in hydro-climatology of the Dinder and Rahad basins, Blue Nile, Ethiopia/Sudan. International Journal of Hydrology Science and Technology 9(6): 690-712.

Hassan, T. A. (2017). Population Estimate 0f Warthog (Phacochoerus aethipicus) in Six Mayas in Dinder National Park (DNP). Poult Fish Wildl Sci 5: 183. Doi: 10.4172/2375-446X.1000183

Hassan, T. A. (2017). Spatial distribution of Hyphaene thebacia and Ziziphus spina Christi in Riverine Forest of Dinder National Park, Sudan. Global Journal of Earth and Environmental Science 2: 15-20.

Hastings, A., J. E. Byers, J. A. Crooks, K. Cuddington, C. G. Jones, J. G. Lambrinos, T. S. Talley and W. G. Wilson (2007). Ecosystem engineering in space and time. Ecology letters 10(2): 153-164.

Hawando, T., 1997. Desertification in Ethiopian highlands. Rala report 200.

Heckmann, T., M. Cavalli, O. Cerdan, S. Foerster, M. Javaux, E. Lode, A. Smetanová, D. Vericat and F. Brardinoni (2018). Indices of sediment connectivity: opportunities, challenges and limitations. Earth-Science Reviews 187: 77-108.

Hemmavanh, C., Y. Ye and A. Yoshida (2010). Forest land use change at Trans-Boundary Laos-China Biodiversity Conservation Area. Journal of Geographical Sciences 20(6): 889-898.

Hessels, T. M. (2015). Comparison and Validation of Several Open Access Remotely Sensed Rainfall Products for the Nile Basin, TU Delft, Delft University of Technology.

Holling, C. S. (1978). Adaptive environmental assessment and management. John Wiley & Sons.

Hu, W.-w., G.-x. Wang, W. Deng and S.-n. Li (2008). The influence of dams on ecohydrological conditions in the Huaihe River basin, China. Ecological Engineering 33(3): 233-241.

Hurni, H., K. Tato and G. Zeleke (2005). The implications of changes in population, land use, and land management for surface runoff in the upper Nile basin area of Ethiopia. Mountain research and development 25(2): 147-154.

Hurst, H., R. Black and Y. Simaika (1959). The Nile Basin, vol. IX. The hydrology of the Blue Nile and Atbara and the Main Nile to Aswan, with reference to some Projects. Ministry of Public Works, Physical Department, Cairo, Egypt.

Ibrahim, Y. A., M. S. R. Elnil and A. A. Ahmed (2009). Improving Water Management Practices in the Rahad Scheme. Improved Water and Land Management in the Ethiopian Highlands: Its Impact on Downstream Stakeholders Dependent on the Blue Nile, Intermediate Results Dissemination Workshop. February 5-6, 2009, Addis Ababa, Ethiopia., IWMI Subregional Office for East Africa and Nile Basin, Addis Ababa, Ethiopia.: 50-69.

Jain, S. K., R. Singh, M. Jain and A. Lohani (2005). Delineation of flood-prone areas using remote sensing techniques. Water Resources Management 19(4): 333-347.

Jensen, J. R. (2005). Digital change detection. Introductory digital image processing: A remote sensing perspective. Prentice Hall, Upper Saddle River, NY, 525 pp.

Jewitt, G. (2002). Can integrated water resources management sustain the provision of ecosystem goods and services? Physics and Chemistry of the Earth, Parts A/B/C 27(11-22): 887-895.

Jones, C. G., J. L. Gutiérrez, J. E. Byers, J. A. Crooks, J. G. Lambrinos and T. S. Talley (2010). A framework for understanding physical ecosystem engineering by organisms. Oikos 119(12): 1862-1869.

Jones, C. G., J. H. Lawton and M. Shachak (1994). Organisms as ecosystem engineers. Oikos: 373-386.

Jones, C. G., J. H. Lawton and M. Shachak (1997). Positive and negative effects of organisms as physical ecosystem engineers. Ecology 78(7): 1946-1957.

Jordan, W. R., M. E. Gilpin and J. D. Aber (1990). Restoration ecology: a synthetic approach to ecological research, Cambridge University Press.

Kahya, E. and S. Kalaycı (2004). Trend analysis of streamflow in Turkey. Journal of Hydrology 289(1): 128-144.

Kamusoko, C. and M. Aniya (2009). Hybrid classification of Landsat data and GIS for land use/cover change analysis of the Bindura district, Zimbabwe. International Journal of Remote Sensing 30(1): 97-115.

Karssenberg, D. (2002). The value of environmental modelling languages for building distributed hydrological models. Hydrological Processes 16(14): 2751-2766.

Keddy, P. A., H. T. Lee and I. C. Wisheu (1993). Choosing indicators of ecosystem integrity: wetlands as a model system. Ecological integrity and the management of ecosystems: 61-82.

Kendall, M. (1975). Rank correlation measures. Charles Griffin, London 202.

Köhler, L., M. Mulligan, J. Schellekens, S. Schmid and C. Tobón (2006). Final Technical Report DFID-FRP Project no. R7991 Hydrological impacts of converting tropical montane cloud forest to pasture, with initial reference to northern Costa Rica.

Kondolf, G. M. (2006). Process-based ecological river restoration: visualizing three-dimensional connectivity and dynamic vectors to recover lost linkages. Ecology and Society 11(2): 1.

Kundzewicz, Z. W. (2003). Ecohydrology for sustainable wetlands under global change–data, models, management. Measurement techniques and data assessment in wetland hydrology: 25-35.

Langeland, K. and K. C. Burks (1999). Identification & biology of non-native plants in Florida's natural areas. University of Florida, Gainesville, Florida.

Lehmann, E. and H. D'abrera (1975). Nonparametrics: Statistical methods based on ranks, holden-day inc. San Francisco: 300-315.

Lesser, G., J. Roelvink, J. Van Kester and G. Stelling (2004). Development and validation of a three-dimensional morphological model. Coastal engineering 51(8): 883-915.

Lettenmaier, D. P., E. F. Wood and J. R. Wallis (1994). Hydro-climatological trends in the continental United States, 1948-88. Journal of Climate 7(4): 586-607.

Lew, S., K. Glińska-Lewczuk, P. Burandt, K. Obolewski, A. Goździejewska, M. Lew and J. Dunalska (2016). Impact of environmental factors on bacterial communities in floodplain lakes differed by hydrological connectivity. Limnologica 58: 20-29.

Li, Y., Q. Zhang, Y. Cai, Z. Tan, H. Wu, X. Liu and J. Yao (2019). Hydrodynamic investigation of surface hydrological connectivity and its effects on the water quality of seasonal lakes: insights from a complex floodplain setting (Poyang Lake, China). Science of The Total Environment 660: 245-259.

Lins, H. F. and J. R. Slack (1999). Streamflow trends in the United States. Geophysical Research Letters 26, No. 2: 227-230.

Lytle, D. A. and N. L. Poff (2004). Adaptation to natural flow regimes. Trends in Ecology & Evolution 19(2): 94-100.

Maddock, I. (1999). The importance of physical habitat assessment for evaluating river health. Freshwater biology 41(2): 373-391.

Mander, M., G. Jewitt, J. Dini, J. Glenday, J. Blignaut, C. Hughes, C. Marais, K. Maze, B. van der Waal and A. Mills (2017). Modelling potential hydrological returns from investing in ecological infrastructure: Case studies from the Baviaanskloof-Tsitsikamma and uMngeni catchments, South Africa. Ecosystem Services 27: 261-271.

Mann, H. B. (1945). Nonparametric tests against trend. Econometrica: Journal of the Econometric Society Vol. 13, No. 3: 245-259.

Marcotullio, P. J. and T. Onishi (2008). The impact of urbanization on soils. Land use and soil resources, Springer: 201-250.

Masih, I., S. Maskey, F. Mussá and P. Trambauer (2014). A review of droughts on the African continent: a geospatial and long-term perspective. Hydrology and Earth System Sciences 18(9): 3635-3649.

McCalin, M. E., L. Chícharo, N. Fohrer, M. G. Novillo, W. Windhorst and M. Zalewski (2012). Training hydrologists to be ecohydrologists and play a leading role in environmental problem solving.

McFeeters, S. K. (1996). The use of the Normalized Difference Water Index (NDWI) in the delineation of open water features. International journal of remote sensing 17(7): 1425-1432.

Mengistu, D., W. Bewket and R. Lal (2014). Recent spatiotemporal temperature and rainfall variability and trends over the Upper Blue Nile River Basin, Ethiopia. International Journal of Climatology 34(7): 2278-2292.

Metzger, M., M. Rounsevell, L. Acosta-Michlik, R. Leemans and D. Schröter (2006). The vulnerability of ecosystem services to land use change. Agriculture, Ecosystems & Environment 114(1): 69-85.

Mohler, C. and M. Liebman (1987). Weed productivity and composition in sole crops and intercrops of barley and field pea. Journal of applied ecology: 685-699.

Mubarak, A. (2010). Census of the wild animals in the Dinder National Park for the year 2010. Wildlife Research Center.

Mukul, M., V. Srivastava and M. Mukul (2015). Analysis of the accuracy of Shuttle Radar Topography Mission (SRTM) height models using International Global Navigation Satellite System Service (IGS) Network. Journal of Earth System Science 124(6): 1343-1357.

Mundia, C. and M. Aniya (2006). Dynamics of landuse/cover changes and degradation of Nairobi City, Kenya. Land Degradation & Development 17(1): 97-108.

Mutasim B, Frazer T (2004) Paper (10) Protected Areas Management. doi:https://studylib.net/doc/7332179/paper--10--protected-areas-management. Accessed 25 June 2019.

Naiman, R. J. and H. Décamps (1997). The ecology of interfaces: riparian zones. Annual review of ecology and systematics: 621-658.

Naiman, R. J., C. A. Johnston and J. C. Kelley (1988). Alteration of North American streams by beaver. BioScience 38(11): 753-762.

Nash, J. E. and J. V. Sutcliffe (1970). River flow forecasting through conceptual models part I- A discussion of principles. Journal of Hydrology 10(3): 282-290.

Nawaz, N., T. Bellerby, M. Sayed and M. Elshamy (2010). Blue Nile runoff sensitivity to climate change. Open Hydrology 4, special issue: 137-151.

Newbury, R. W. (1984). Hydrologic determinants of aquatic insect habitats.

Noss, R. F. (1990). Indicators for monitoring biodiversity: a hierarchical approach. Conservation Biology 4(4): 355-364.

Nowell, A. and P. Jumars (1984). Flow environments of aquatic benthos. Annual review of ecology and systematics: 303-328.

Nyssen, J., J. Pontzeele and P. Billi (2011). Effect of beaver dams on the hydrology of small mountain streams: example from the Chevral in the Ourthe Orientale basin, Ardennes, Belgium. Journal of Hydrology 402(1-2): 92-102.

Ockenden, M. C., C. Deasy, J. N. Quinton, A. P. Bailey, B. Surridge and C. Stoate (2012). Evaluation of field wetlands for mitigation of diffuse pollution from agriculture: sediment retention, cost and effectiveness. Environmental Science & Policy 24: 110-119.

Onyutha, C. (2016). Identification of sub-trends from hydro-meteorological series. Stochastic environmental research and risk assessment 30(1): 189-205.

Palmer, M. A., H. L. Menninger and E. Bernhardt (2010). River restoration, habitat heterogeneity and biodiversity: a failure of theory or practice? Freshwater Biology 55: 205-222.

Partal, T. and E. Kahya (2006). Trend analysis in Turkish precipitation data. Hydrological processes 20(9): 2011-2026.

Partheniades, E. (1965). Erosion and deposition of cohesive soils. Journal of the Hydraulics Division 91(1): 105-139.

Patnaik, S. (1976). Autecology of Ipomoea aquatica Forsk. J. Inland Fish. Soc. India 8: 77-82.

Petersen, G., J. Abeya and N. Fohrer (2007). Spatio-temporal water body and vegetation changes in the Nile swamps of southern Sudan. Advances in Geosciences 11: 113-116.

Pettitt, A. (1979). A non-parametric approach to the change-point problem. Applied statistics 28(2): 126-135.

Poff, N. L., J. D. Allan, M. B. Bain, J. R. Karr, K. L. Prestegaard, B. D. Richter, R. E. Sparks and J. C. Stromberg (1997). The natural flow regime. BioScience 47(11): 769-784.

Polasky, S., E. Nelson, D. Pennington and K. Johnson (2011). The Impact of Land-Use Change on Ecosystem Services, Biodiversity and Returns to Landowners: A Case Study in the State of Minnesota. Environmental and Resource Economics 48(2): 219-242.

Poole, G. C. (2002). Fluvial landscape ecology: addressing uniqueness within the river discontinuum. Freshwater biology 47(4): 641-660.

Popescu, I., E. Cioaca, Q. Pan, A. Jonoski and J. Hanganu (2015). Use of hydrodynamic models for the management of the Danube Delta wetlands: The case study of Sontea-Fortuna ecosystem. Environmental Science & Policy 46: 48-56.

Pradhan, B. and Z. Suleiman (2009). Landcover mapping and spectral analysis using multi-sensor satellite data fusion techniques: case study in Tioman Island, Malaysia. Journal of Geomatics 3(2): 71-78.

Rebelo, L.-M., R. Johnston, T. Hein, G. Weigelhofer, T. D'Haeyer, B. Kone and J. Cools (2012). Challenges to the integration of wetlands into IWRM: The case of the Inner Niger Delta (Mali) and the Lobau Floodplain (Austria). Environmental Science & Policy 34: 58-68.

Richards, J., J. Xiuping, W. Gessner and D. Ricken (2006). Remote sensing digital image analysis. 4th edition-Springer-Verlag, Berlin Heidelberg: ISBN: 13 978-973-540-25128-25126. pp.25194 – 25199.

Richter, B., J. Baumgartner, R. Wigington and D. Braun (1997). How much water does a river need? Freshwater biology 37(1): 231-249.

Richter, B. D., J. V. Baumgartner, J. Powell and D. P. Braun (1996). A method for assessing hydrologic alteration within ecosystems. Conservation Biology 10(4): 1163-1174.

Rientjes, T., A. Haile, E. Kebede, C. Mannaerts, E. Habib and T. Steenhuis (2011). Changes in land cover, rainfall and stream flow in Upper Gilgel Abbay catchment, Blue Nile basin–Ethiopia. Hydrology and Earth System Sciences 15(6): 1979-1989.

Robinson, C., K. Tockner and J. Ward (2002). The fauna of dynamic riverine landscapes. Freshwater biology 47(4): 661-677.

Rodríguez-Iturbe, I. and A. Porporato (2005). Ecohydrology of water-controlled ecosystems: soil moisture and plant dynamics, Cambridge University Press.

Rodriguez-Puebla, C., A. Encinas, S. Nieto and J. Garmendia (1998). Spatial and temporal patterns of annual precipitation variability over the Iberian Peninsula. International Journal of Climatology 18(3): 299-316.

RRP (1993). Phase I Feasibility Study: Final Report. The River Restoration Project, Huntingdon, UK.

Santisteban, J. I., R. Mediavilla, L. G. de Frutos and I. L. Cilla (2019). Holocene floods in a complex fluvial wetland in central Spain: Environmental variability, climate and time. Global and Planetary Change 181: 102986.

Savenije, H. H. and P. Van der Zaag (2008). Integrated water resources management: Concepts and issues. Physics and Chemistry of the Earth, Parts A/B/C 33(5): 290-297.

Schellekens, J. (2011). WFlow, a flexible hydrological model, available at: https://publicwiki.deltares.nl/download/attachments/33226762/wflow.pdf?version=1.

Schröder, B. (2006). Pattern, process, and function in landscape ecology and catchment hydrology? how can quantitative landscape ecology support predictions in ungauged basins? Hydrology and Earth System Sciences Discussions 10(6): 967-979.

Scott, M. L., G. T. Auble and J. M. Friedman (1997). Flood dependency of cottonwood establishment along the Missouri River, Montana, USA. Ecological Applications 7(2): 677-690.

Sethre, P. R., B. C. Rundquist and P. E. Todhunter (2005). Remote detection of prairie pothole ponds in the Devils Lake Basin, North Dakota. GIScience & Remote Sensing 42(4): 277-296.

Shaltout, K., H. El-Kady and Y. Al-Sodany (1995). Vegetation analysis of the Mediterranean region of Nile Delta. Vegetatio 116(1): 73-83.

Shaltout, K. H., L. M. Hassan and E. A. Farahat (2005). Vegetation-environment relationships in south Nile Delta. Taeckholmia. 2005a 25: 15-46.

Shang, H., J. Yan, M. Gebremichael and S. M. Ayalew (2011). Trend analysis of extreme precipitation in the Northwestern Highlands of Ethiopia with a case study of Debre Markos. Hydrology and Earth System Sciences 15(6): 1937.

Shankman, D. (1996). Stream Channelization and Changing Vegetation Patterns in the U. S. Coastal Plain. Geographical Review 86(2): 216-232.

Sivapalan, M. (2005). Pattern, process and function: elements of a unified theory of hydrology at the catchment scale. Encyclopedia of hydrological sciences.

Sneyers, R. (1990). On the statistical analysis of series of observations.

Spearman, C. (1904). The proof and measurement of association between two things. The American journal of psychology 15, No. 1(1): 72-101.

Stafford, J., G. Wendler and J. Curtis (2000). Temperature and precipitation of Alaska: 50 year trend analysis. Theoretical and Applied Climatology 67(1): 33-44.

Stanford, J. A. (2002). Rivers in the landscape: introduction to the special issue on riparian and groundwater ecology. Freshwater biology 40(3): 402-406.

Sutcliffe, J. V. and Y. P. Parks (1999). The hydrology of the Nile, International Association of Hydrological Sciences Wallingford, Oxfordshire, UK.

Tabacchi, E., D. L. Correll, R. Hauer, G. Pinay, A. M. Planty-Tabacchi and R. C. Wissmar (1998). Development, maintenance and role of riparian vegetation in the river landscape. Freshwater biology 40(3): 497-516.

Tabacchi, E., L. Lambs, H. Guilloy, A.-M. Planty-Tabacchi, E. Muller and H. Decamps (2000). Impacts of riparian vegetation on hydrological processes. Hydrological Processes 14(16-17): 2959-2976.

Tan, Z., Y. Li, X. Xu, J. Yao and Q. Zhang (2019). Mapping inundation dynamics in a heterogeneous floodplain: Insights from integrating observations and modeling approach. Journal of Hydrology 572: 148-159.

Teferi, E., W. Bewket, S. Uhlenbrook and J. Wenninger (2013). Understanding recent land use and land cover dynamics in the source region of the Upper Blue Nile, Ethiopia: Spatially explicit statistical modeling of systematic transitions. Agriculture, Ecosystems & Environment 165(0): 98-117.

Teferi, E., S. Uhlenbrook, W. Bewket, J. Wenninger and B. Simane (2010). The use of remote sensing to quantify wetland loss in the Choke Mountain range, Upper Blue Nile basin, Ethiopia. Hydrology and Earth System Sciences 14(12): 2415-2428.

Tekleab, S., Y. Mohamed and S. Uhlenbrook (2013). Hydro-climatic trends in the Abay/Upper Blue Nile basin, Ethiopia. Physics and Chemistry of the Earth, Parts A/B/C 61–62: 32-42.

Tesemma, Z. K., Y. A. Mohamed and T. S. Steenhuis (2010). Trends in rainfall and runoff in the Blue Nile Basin: 1964–2003. Hydrological processes 24(25): 3747-3758.

The Nature Conservancy (2009). Indicators of Hydrologic Alteration Version 7.1 User's Manual, The Nature Conservancy. Virginia, United States.

Tockner, K., J. Ward, P. Edwards and J. Kollmann (2002). Riverine landscapes: an introduction. Freshwater biology 47(4): 497-500.

Turner, B. L., E. F. Lambin and A. Reenberg (2007). The emergence of land change science for global environmental change and sustainability. Proceedings of the National Academy of Sciences 104(52): 20666-20671.

Turner, B. L., P. A. Matson, J. J. McCarthy, R. W. Corell, L. Christensen, N. Eckley, G. K. Hovelsrud-Broda, J. X. Kasperson, R. E. Kasperson and A. Luers (2003). Illustrating the coupled human–environment system for vulnerability analysis: three case studies. Proceedings of the National Academy of Sciences 100(14): 8080-8085.

Turner, M. G. (2005). Landscape ecology: what is the state of the science? Annual Review of Ecology, Evolution, and Systematics: 319-344.

Uhlenbrook, S. (2007). Biofuel and water cycle dynamics: what are the related challenges for hydrological processes research? Hydrological processes 21(26): 3647-3650.

Uhlenbrook, S., S. Roser and N. Tilch (2004). Hydrological process representation at the meso-scale: the potential of a distributed, conceptual catchment model. Journal of Hydrology 291(3): 278-296.

Van Rijn, L. C. (1984). Sediment transport, part II: suspended load transport. Journal of hydraulic engineering 110(11): 1613-1641.

Vertessy, R. A. and H. Elsenbeer (1999). Distributed modeling of storm flow generation in an Amazonian rain forest catchment: Effects of model parameterization. Water Resources Research 35(7): 2173-2187.

Vinnikov, K. Y. and N. C. Grody (2003). Global warming trend of mean tropospheric temperature observed by satellites. Science 302(5643): 269-272.

Warburton, M. L., R. E. Schulze and G. P. Jewitt (2012). Hydrological impacts of land use change in three diverse South African catchments. Journal of Hydrology 414: 118-135.

Ward, J., K. Tockner, D. Arscott and C. Claret (2002). Riverine landscape diversity. Freshwater Biology 47(4): 517-539.

Wardrop, D. H., M. E. Kentula, S. F. Jensen, D. L. Stevens, K. C. Hychka and R. P. Brooks (2007). Assessment of wetlands in the Upper Juniata watershed in Pennsylvania, USA using the hydrogeomorphic approach. Wetlands 27(3): 432-445.

Wassen, M. J. and A. P. Grootjans (1996). Ecohydrology: an interdisciplinary approach for wetland management and restoration. Plant Ecology 126(1): 1-4.

Weilhoefer, C. (2011). A review of indicators of estuarine tidal wetland condition. Ecological indicators 11(2): 514-525.

Wiens, J. A. (2002). Riverine landscapes: taking landscape ecology into the water. Freshwater biology 47(4): 501-515.

Wilson, M., P. Bates, D. Alsdorf, B. Forsberg, M. Horritt, J. Melack, F. Frappart and J. Famiglietti (2007). Modeling large-scale inundation of Amazonian seasonally flooded wetlands. Geophysical Research Letters 34(15).

Wohl, E., P. L. Angermeier, B. Bledsoe, G. M. Kondolf, L. MacDonnell, D. M. Merritt, M. A. Palmer, N. L. Poff and D. Tarboton (2005). River restoration. Water Resources Research 41(10).

Woo, M. K. and R. Thorne (2003). Streamflow in the Mackenzie basin, Canada. Arctic: 328-340.

Wood, P. J., D. M. Hannah and J. P. Sadler (2008). Hydroecology and ecohydrology: past, present and future, John Wiley & Sons.

Woolf, D., S. Jirka, E. Milne, M. Easter, S. DeGloria, D. Solomon and J. Lehmann (2015). Climate Change Mitigation Potential of Ethiopia's Productive Safety-Net Program (PSNP).

Wright, J. P. (2009). Linking populations to landscapes: richness scenarios resulting from changes in the dynamics of an ecosystem engineer. Ecology 90(12): 3418-3429.

Xu, H. (2006). Modification of normalised difference water index (NDWI) to enhance open water features in remotely sensed imagery. International journal of remote sensing 27(14): 3025-3033.

Young, P. C. and M. Ratto (2009). A unified approach to environmental systems modeling. Stochastic Environmental Research and Risk Assessment 23(7): 1037-1057.

Yousif, R. A. and F. A. Mohamed (2012). Trends of poaching, Livestock Trespassing, Fishing and Resource Collection from 1986-2010 in Dinder National Park, Sudan. J. Life Sci. Biomed 2(3): 105-110.

Yu, X., J. Hawley-Howard, A. L. Pitt, J.-J. Wang, R. F. Baldwin and A. T. Chow (2015). Water quality of small seasonal wetlands in the Piedmont ecoregion, South Carolina, USA: effects of land use and hydrological connectivity. water research 73: 98-108.

Yuan, F., M. E. Bauer, N. J. Heinert and G. R. Holden (2005). Multi-level land cover mapping of the Twin Cities (Minnesota) metropolitan area with multi-seasonal Landsat TM/ETM+ data. Geocarto International 20(2): 5-13.

Yue, S., P. J. Pilon, B. Phinney and G. Cavadias (2002). The influence of autocorrelation on the ability to detect trend in hydrological series. Hydrol. Processes 16(9), 1807–1829.

Zalewski, M. (2002). Ecohydrology-The use of ecological and hydrological processes for sustainable management of water resources/Ecohydrologie-La prise en compte de processus écologiques et hydrologiques pour la gestion durable des ressources en eau. Hydrological Sciences Journal 47(5): 823-832.

Zalewski, M., M. McClain and S. Eslamian (2016). New challenges and dimensions of Ecohydrology-enhancement of catchments sustainability potential. Ecohydrology & Hydrobiology 16(1): 1-3.

Zedler, J. B. and S. Kercher (2005). Wetland resources: status, trends, ecosystem services, and restorability. Annu. Rev. Environ. Resour. 30: 39-74.

Zeleke, G. and H. Hurni (2001). Implications of land use and land cover dynamics for mountain resource degradation in the northwestern Ethiopian highlands. Mountain research and development 21(2): 184-191.

Zeug, S. and K. Winemiller (2008). Relationships between hydrology, spatial heterogeneity, and fish recruitment dynamics in a temperate floodplain river. River Research and Applications 24(1): 90-102.

# APPENDICES

**Appendix A:** ***The Wflow_sbm interception model*:**

The analytical model of rainfall interception is based on Rutter's numerical model (see Gash, 1979; Gash et al., 1995, for a full description). The simplifications that Gash (1979) introduced allow the model to be applied on a daily basis. The amount of water needed to completely saturate the canopy ($P'$) is defined as:

$$P' = \frac{-\bar{R}S}{\bar{E}_w} ln \left[1 - \frac{\bar{E}_w}{\bar{R}}(1 - p - p_t)^{-1}\right] \qquad \text{(A1)}$$

where:

$\bar{R}$ = average precipitation on a saturated canopy [mm day$^{-1}$]

$\bar{E}_w$ = average evaporation from the wet canopy [mm day$^{-1}$]

$S$ = canopy storage capacity [mm]

$p$ = free throughfall coefficient: the proportion of rain which falls to the ground without sticking the canopy [-]

$p_t$ = proportion of rain that is diverted to stemflow [-]

Interception losses from the stems are calculated for days with $P \geq S_t/P_t$. $S_t$ (trunk water capacity [mm]) and $P_t$ are small and neglected in the wflow_sbm model. In applying the analytical model, saturated conditions are assumed to occur when the hourly rainfall exceeds a certain threshold. Often a threshold of 0.5 mm hr$^{-1}$ is used (Gash, 1979). $\bar{R}$ is calculated for all hours when the rainfall exceeds the threshold to give an estimate of the mean rainfall rate onto a saturated canopy. $E_w$ is then calculated using the Rutter model.

***The wflow_sbm soil water accounting scheme:***

Within the soil model, the soil is considered as a bucket with a certain depth ($Z_t$), divided into a saturated store (S) and an unsaturated store (U), the capacity of each is expressed in units of depth (mm). The top of the saturated store forms a pseudo-water table at depth ($Z_i$) such that the value of (S) at any time is given by:

$$S = (z_t - z_i)(\theta_s - \theta_r) \qquad \text{(A2)}$$

Where:

$\theta_s$ and $\theta_r$ (mm) are the saturated and residual soil water contents, respectively.

The unsaturated store (U) is subdivided into storage ($U_s$) and deficit ($U_d$) which are also expressed in units of depth:

$$U_d = (\theta_s - \theta_r)z_i - U \qquad \text{(A3)}$$

and

$$U_s = U - U_d \qquad \text{(A4)}$$

The saturation deficit ($S_d$) for the whole soil profile is defined as:

$$S_d = (\theta_s - \theta_r)z_t - S \qquad \text{(A5)}$$

Infiltrating rainfall enters the unsturated store first. The transfer of water from the unsaturated store to the saturated store (st) is controlled by the saturated hydraulic conductivity $K_{sat}$ at depth ($Z_i$) and the ratio between $U_s$ and $S_d$.

$$st = K_{sat}\frac{U_s}{S_d} \qquad \text{(A6)}$$

As the saturation deficit becomes smaller, the rate of the transfer between the unsaturated and saturated stores increases. Saturated conductivity ($K_{sat}$) declines with soil depth (z) in the model according to:

$$K_{sat} = K_0\, e^{(-fz)} \qquad \text{(A7)}$$

where:

$K_0$ is the saturated conductivity at the soil surface [m day$^{-1}$] and;

$f$ is a scaling parameter [m$^{-1}$]

The scaling parameter f is defined by:

$$f = \frac{\theta_s - \theta_r}{M} \qquad \text{(A8)}$$

M is a soil parameter determining the decrease of saturated conductivity with depth [m].

The saturated store can be drained laterally via subsurface flow according to:

$$sf = K_0 tan\,(\beta)\, e^{-S_d/M} \qquad \text{(A9)}$$

where:

β is element slope angle [deg.]

$sf$ is the calculated subsurface flow [$m^2$ $day^{-1}$]

The original SBM model does not include transpiration or a notion of capillary rise. In wflow_sbm transpiration is first taken from the saturated store if the roots reach the water table ($Z_i$). If the saturated store cannot satisfy the demand the unsaturated store is used next. First the number of wet roots (WR) is determined (going from 1 to 0) using a sigmoid function as follows:

$$WR = \;1.0/(1.0 + e^{-SN(WT-RT)}) \quad \text{(A10)}$$

where:

SN is sharpness parameters [-]

WT is water table [mm]

RT is rooting depth [mm]

The sharpness parameter (by default a large negative value, estimated as -80000) is a parameter that determines if there is a stepwise output or a more gradual output (default is stepwise). Water Table is the level of the water table in the grid cell below the surface and rooting depth is the maximum depth of the roots below the surface. For all values of water tables smaller that rooting depth a value of 1 is returned, if they are equal to rooting depth a value of 0.5 is returned, and if the water table is larger than the rooting depth a value of zero is returned. The returned wet roots (WR) fraction is multiplied by the potential evaporation (and limited by the available water in saturated store) to get the transpiration from the saturated part of the soil. Next the remaining potential evaporation is used to extract water from the unsaturated store.

Capillary rise is determined using the following approach: first the $K_{sat}$ is determined at the water table ($Z_i$); next a potential capillary rise is determined from the minimum of the $K_{sat}$, the actual transpiration taken from the unsaturated store, the available water in the saturated store and the deficit of the unsaturated store. Finally, the potential rise is scaled using the distance between the roots and the water table using:

$$CS = CSF/(CSF + z_i - RT) \quad (A11)$$

in which CS is the scaling factor to multiply the potential rise with, CSF is a model parameter (default = 100) and RT is the rooting depth. If the roots reach the water table (RT > Zi) CS is set to zero and thus setting the capillary rise to zero. A detailed description of the TOPOG_SBM model has been provided by Vertessy and Elsenbeer (1999).

## Appendix B: Wflow model parameter's description:

| Parameter name in Wflow | Description | Unit |
| --- | --- | --- |
| CanopyGapFraction | Gash interception model parameter: the free throughfall coefficient. Fraction of precipitation that does not hit the canopy directly | [-] |
| EoverR (E/R) | Gash interception model parameter. Ratio of average wet canopy evaporation rate over average precipitation rate. | [-] |
| MaxCanopyStorage | Canopy storage. Used in the Gash interception model | [mm] |
| FirstZoneCapacity | Maximum capacity of the saturated store. | [mm] |
| FirstZoneKsatVer | Saturated conductivity of the store at the surface. The M parameter determines how this decrease with depth. | [mm] |
| FirstZoneMinCapacity | Minimum capacity of the saturated store [mm] | [mm] |
| InfiltCapPath | Infiltration capacity of the compacted soil fraction of each grid cell. | [mm/day] |
| InfiltCapSoil | Infiltration capacity of the non-compacted soil fraction of each grid cell | [mm/day] |
| M | Soil parameter determining the decrease of saturated conductivity with depth. | [m] |
| N | Manning N parameter for the Kinematic wave function. | |
| N_river | Manning's parameter for cells marked as river | |
| LeafAreaIndex | Total one-side green leaf area per ground surface area. | [-] |
| Albedo | Reflectivity of earth surface: the ratio of radiation reflected to the radiation incident on a surface. | [-] |
| Beta | element slope angle | [degree] |
| rootdistpar | Sharpness parameter determine how roots are linked to water table. | [mm] |
| PathFrac | Fraction of compacted area per grid cell. | [-] |
| RootingDepth | Rooting depth of the vegetation. | [mm] |
| CapScale | Scaling factor in the Capillary rise calculations | [mm/day] |
| RunoffGeneratingGWPerc | Fraction of the soil depth that contributes to sub-cell runoff | [-] |
| thetaR | Residual water content. | [-] |
| thetaS | Water content at saturation (porosity). | [-] |

## Appendix C: Wflow model parameters calibrated values:

Albedo

| Land cover | Sub-catchment | Soil type | Value |
|---|---|---|---|
| 1 | [0,> | [0,> | 0.40 |
| 2 | [0,> | [0,> | 0.20 |
| 3 | [0,> | [0,> | 0.16 |
| 4 | [0,> | [0,> | 0.26 |
| 5 | [0,> | [0,> | 0.25 |
| 6 | [0,> | [0,> | 0.10 |

CanopyGapFraction

| Land cover | Sub-catchment | Soil type | Value |
|---|---|---|---|
| 1 | [0,> | [0,> | 1.0 |
| 2 | [0,> | [0,> | 0.2 |
| 3 | [0,> | [0,> | 0.6 |
| 4 | [0,> | [0,> | 0.5 |
| 5 | [0,> | [0,> | 0.4 |
| 6 | [0,> | [0,> | 0.5 |

EoverR

| Land cover | Sub-catchment | Soil type | Value |
|---|---|---|---|
| 1 | [0,> | [0,> | 0.0 |
| 2 | [0,> | [0,> | 0.3 |
| 3 | [0,> | [0,> | 0.2 |
| 4 | [0,> | [0,> | 0.2 |
| 5 | [0,> | [0,> | 0.1 |
| 6 | [0,> | [0,> | 0.0 |

FirstZoneCapacity

| Land cover | Sub-catchment | Soil type | Value |
|---|---|---|---|
| [0,> | [0,> | 1 | 44500 |
| [0,> | [0,> | 2 | 42000 |
| [0,> | [0,> | 3 | 44500 |
| [0,> | [0,> | 4 | 39000 |
| [0,> | [0,> | 5 | 44000 |
| [0,> | [0,> | 6 | 42000 |
| [0,> | [0,> | 7 | 44500 |

FirstZoneKsatVer

| Land cover | Sub-catchment | Soil type | Value |
|---|---|---|---|
| [0,> | [0,> | 1 | 511 |

FirstZoneMinCapacity

| Land cover | Sub-catchment | Soil type | Value |
|---|---|---|---|

| | | | |
|---|---|---|---|
| [0,> | [0,> | 2 | 600 |
| [0,> | [0,> | 3 | 543 |
| [0,> | [0,> | 4 | 525 |
| [0,> | [0,> | 5 | 586 |
| [0,> | [0,> | 6 | 576 |
| [0,> | [0,> | 7 | 540 |

InfiltCapPath

| Land cover | Sub-catchment | Soil type | Value |
|---|---|---|---|
| [0,> | [0,> | 1 | 5 |
| [0,> | [0,> | 2 | 21 |
| [0,> | [0,> | 3 | 5 |
| [0,> | [0,> | 4 | 32 |
| [0,> | [0,> | 5 | 34 |
| [0,> | [0,> | 6 | 5 |
| [0,> | [0,> | 7 | 21 |

LeafAreaIndex

| Land cover | Sub-catchment | Soil type | Value |
|---|---|---|---|
| 1 | [0,> | [0,> | 0.0 |
| 2 | [0,> | [0,> | 8.8 |
| 3 | [0,> | [0,> | 7.0 |
| 4 | [0,> | [0,> | 0.6 |

| | | | |
|---|---|---|---|
| [0,> | [0,> | 1 | 125 |
| [0,> | [0,> | 2 | 50 |
| [0,> | [0,> | 3 | 137.5 |
| [0,> | [0,> | 4 | 33 |
| [0,> | [0,> | 5 | 87.5 |
| [0,> | [0,> | 6 | 60 |
| [0,> | [0,> | 7 | 70 |

InfiltCapSoil

| Land cover | Sub-catchment | Soil type | Value |
|---|---|---|---|
| [0,> | [0,> | 1 | 24 |
| [0,> | [0,> | 2 | 103 |
| [0,> | [0,> | 3 | 24 |
| [0,> | [0,> | 4 | 158 |
| [0,> | [0,> | 5 | 170 |
| [0,> | [0,> | 6 | 100 |
| [0,> | [0,> | 7 | 103 |

M

| Land cover | Sub-catchment | Soil type | Value |
|---|---|---|---|
| [0,> | [0,> | 1 | 100 |
| [0,> | [0,> | 2 | 87 |
| [0,> | [0,> | 3 | 100 |

| | | | |
|---|---|---|---|
| 5 | [0,> | [0,> | 0.7 |
| 6 | [0,> | [0,> | 0.0 |

MaxCanopyStorage

| Land cover | Sub-catchment | Soil type | Value |
|---|---|---|---|
| 1 | [0,> | [0,> | 0.00 |
| 2 | [0,> | [0,> | 0.336 |
| 3 | [0,> | [0,> | 0.21 |
| 4 | [0,> | [0,> | 0.25 |
| 5 | [0,> | [0,> | 0.34 |
| 6 | [0,> | [0,> | 0.00 |

PathFrac

| Land cover | Sub-catchment | Soil type | Value |
|---|---|---|---|
| [0,> | [0,> | 1 | 0.06 |
| [0,> | [0,> | 2 | 0.09 |
| [0,> | [0,> | 3 | 0.05 |
| [0,> | [0,> | 4 | 0.06 |
| [0,> | [0,> | 5 | 0.06 |
| [0,> | [0,> | 6 | 0.07 |
| [0,> | [0,> | 7 | 0.08 |

| | | | |
|---|---|---|---|
| [0,> | [0,> | 4 | 77 |
| [0,> | [0,> | 5 | 100 |
| [0,> | [0,> | 6 | 100 |
| [0,> | [0,> | 7 | 100 |

N

| Land cover | Sub-catchment | Soil type | Value |
|---|---|---|---|
| 1 | [0,> | [0,> | 0.42 |
| 2 | [0,> | [0,> | 0.80 |
| 3 | [0,> | [0,> | 0.70 |
| 4 | [0,> | [0,> | 0.65 |
| 5 | [0,> | [0,> | 0.80 |
| 6 | [0,> | [0,> | 0.12 |

RootingDepth

| Land cover | Sub-catchment | Soil type | Value |
|---|---|---|---|
| 1 | [0,> | [0,> | 1000 |
| 2 | [0,> | [0,> | 1800 |
| 3 | [0,> | [0,> | 1400 |
| 4 | [0,> | [0,> | 1600 |
| 5 | [0,> | [0,> | 200 |
| 6 | [0,> | [0,> | 0 |

thetaR

| Land cover | Sub-catchment | Soil type | Value |
|---|---|---|---|
| [0,> | [0,> | 1 | 0.15 |
| [0,> | [0,> | 2 | 0.09 |
| [0,> | [0,> | 3 | 0.19 |
| [0,> | [0,> | 4 | 0.09 |
| [0,> | [0,> | 5 | 0.11 |
| [0,> | [0,> | 6 | 0.09 |
| [0,> | [0,> | 7 | 0.08 |

thetaS

| Land cover | Sub-catchment | Soil type | Value |
|---|---|---|---|
| [0,> | [0,> | 1 | 0.5 |
| [0,> | [0,> | 2 | 0.2 |
| [0,> | [0,> | 3 | 0.5 |
| [0,> | [0,> | 4 | 0.3 |
| [0,> | [0,> | 5 | 0.4 |
| [0,> | [0,> | 6 | 0.2 |
| [0,> | [0,> | 7 | 0.2 |

RunoffGeneratingGWPerc

| Land cover | Sub-catchment | Soil type | Value |
|---|---|---|---|
| [0,> | [0,> | [0,> | 0.1 |

rootdistpar

| Land cover | Sub-catchment | Soil type | Value |
|---|---|---|---|
| [0,> | [0,> | [0,> | -80000 |

N_River

| Land cover | Sub-catchment | Soil type | Value |
|---|---|---|---|
| [0,> | [0,> | [0,> | 0.035 |

Beta

| Land cover | Sub-catchment | Soil type | Value |
|---|---|---|---|
| [0,> | [0,> | [0,> | 0.6 |

CapScale

| Land cover | Sub-catchment | Soil type | Value |
|---|---|---|---|
| [0,> | [0,> | [0,> | 100 |

Land cover: 1= Bare land, 2= woodland, 3= shrubland, 4= grassland, 5= cropland, 6= water bodies.

Soil type: 1= Vertisols, 2= Luvisols, 3= Nitisols, 4= Leptosols, 5= cambisols, 6= Alisols, 7= Fluvisols.

# LIST OF ACRONYMS

| | |
|---|---|
| AET | Actual Evapotranspiration |
| ANP | Alatish National Park |
| CFL | Courant-Friedrichs-Lewy number |
| CHG | Climate Hazards Group |
| CHIRPS | Climate Hazards Group InfraRed Precipitation with Stations |
| CHPClim | Climate Hazards Precipitation Climatology |
| CSI | Consortium for Spatial Information |
| CGIAR | Consultative Group for International Agricultural Research |
| $D_{50}$ | Median Diameter |
| 3D | Three Dimensional |
| DNP | Dinder National Park |
| DSG | Dominant Soil Group |
| D&R | Dinder and Rahad |
| ET | Evapotranspiration |
| FAO | Food and Agriculutre Organization of the United Nations |
| FEWS | Famine Early Warning System |
| GEO | Group on Earth Observations |
| GIS | Geographic Information System |
| GPS | Glopal positioning System |
| HWSD | Harmonized World Soil Database |
| IHA | Indicators of Hydrologic Alterations |

| | |
|---|---|
| IR | InfraRed |
| JAXA | Japan Aerospace Exploration Agency |
| JD | Julian Date |
| LDD | Local Drainage Direction |
| LULC | Land Use and Land Cover |
| MK | Mann-Kendall |
| MLC | Maximum Likelihood Classification |
| MSS | Multispectral Scanner |
| NASA | National Aeronautics and Space Administration |
| NDVI | Normalized Difference Vegitation Index |
| NDWI | Normalized Difference Water Index |
| NIR | Near InfraRed |
| NSE | Nash–Sutcliffe efficiency |
| PCC | Post-Classification Comparison |
| PET | Potential Evapotranspiration |
| PR | Precipitation Radar |
| RFE | Rainfall Estimates |
| RMSE | Root Mean Square Error |
| RVA | Range of Variability Approach |
| SBRE | Satellites-Based Rainfall Estimates |
| SMU | Soil Mapping Unit |

| | |
|---|---|
| SPP | Species Pluralis, Latin For Multiple Species |
| SRQ | Systematic-Random Quadrat |
| SRTM | Shuttle Radar Topographic Mission |
| TM | Thematic Mapper |
| TMI | TRMM Microwave Image |
| USGS | United States Geological Survey |
| UTM | Universal Transverse Mercator |
| VIRS | Visible and Infrared Scanner |
| WLMP | Water Level Monitoring Points |

# LIST OF TABLES

# LIST OF FIGURES

# ABOUT THE AUTHOR

Khalid E. A. Hassaballah is a civil engineer (Grade: V. Good Honors 2nd Div.1, University of Sinnar, Sudan 2002) with nearly seventeen years of experience covering different aspects of water science and engineering including: Hydroinformatics, catchment hydrology and modelling, hydraulics and river engineering, river morphology, reservoir simulation and optimization, reservoirs bathymetric survey, water resources planning & management and GIS & Remote Sensing applications. He has been working as a researcher at HRC-Sudan since November 2003. In 2010, Eng. Khalid graduated from the MSc Program in Water Science and Engineering, Hydroinformatics specialization, from UNESCO-IHE Institute for Water Education, Delft, The Netherlands. Since November 2012, he has been working on his PhD at the same institute (now IHE-Delft). During the course of his PhD, he studied the impacts of land degradation on the Dinder and Rahad hydrology and morphology and linkages to the ecohydrological system of the Dinder National Park, Sudan, using a combination of different methods. Moreover, he attended the course requirement for the graduate school for Socio-economic and Natural Sciences of the Environment (SENSE).

Between 2010 and 2012, he worked for the Eastern Nile Watershed Management Project (ENWMP) as a water resource specialist for the Sudan component. Beside his PhD work, Khalid has also been involved in many projects at national, regional and international levels. In 2013y, he was involved in the sedimentation and operation study for upper Atbara dam complex (SOSADC) national project as a hydrological modeler. In 2015, he was involved in the Tekezi Atbara project for IGAD as a catchment modelling advisor and supervisor. Later in 2018, Khalid participated as a team member of the Eastern Nile flood preparedness and early warning system at ENTRO. All together has qualified him to be appointed as National team member for the negotiation on filling and operation of Grand Ethiopian Renascence Dam (GERD) since September 2019, and Director for Training and Capacity Building for the Ministry of Irrigation and Water Resources-Sudan since February 2020.

Eng. Khalid is a full member of the Sudan Engineering Society (SES), Sudan Engineering Council (SEC), National Member of Nile Basin Initiative Decision Support System Network (NBI/DSSN) and representative of the IHE-Delft Alumni Sudan. Khalid is also a visiting lecturer at University of Gezira, Civil Engineering Department, and a regularly invited reviewer for several peer-reviewed journals.

**Peer-reviewed journal publications**


1. **Hassaballah, K.**, Y. Mohamed, A. Omer & S. Uhlenbrook, 2020. Modelling the Inundation and Morphology of the Seasonally Flooded Mayas Wetlands in the Dinder National Park-Sudan. Environmental Processes:1-25.
2. T.G. Gebremicael, Y.A. Mohamed, P. van der Zaag, **Khalid Hassaballah**, and E.Y. Hagos. Change in low flows due to catchment management dynamics - application of a comparative modelling approach, Hydrological Processes, 2020.
3. **Hassaballah, K.**, McClain, M., Abdelhameed, S., Mohamed, Y., and Uhlenbrook, S.: Assessing the ecohydrology of the maya wetlands in the Dinder National Park-Sudan (submitted to Ecohydrology and Hydrobiology).
4. **Hassaballah K**, Mohamed YA, Uhlenbrook S (2019) The long-term trends in hydro-climatology of the Dinder and Rahad basins, Blue Nile, Ethiopia/Sudan. International Journal of Hydrology Science and Technology 9:690-712 doi:10.1504/IJHST.2019.103447.
5. **Hassaballah, K.**, Mohamed, Y., Uhlenbrook, S., and Biro, K.: Analysis of streamflow response to land use and land cover changes using satellite data and hydrological modelling: case study of Dinder and Rahad tributaries of the Blue Nile (Ethiopia–Sudan), Hydrol. Earth Syst. Sci., 21, 5217-5242, https://doi.org/10.5194/hess-21-5217-2017, 2017.
6. **Hassaballah, K.**, Jonoski, A., Popescu, I., Solomatine, D.: Model-Based Optimization of Downstream Impact during Filling of a New Reservoir: Case Study of Mandaya/Roseires Reservoirs on the Blue Nile River. Water Resources Management, 1-21, 2011.


**Book chapters:**


7. **Hassaballah, K.**, Y. A. Mohamed and S. Uhlenbrook.: The Mayas wetlands of the Dinder and Rahad: tributaries of the Blue Nile Basin (Sudan). The Wetland Book: II: Distribution, Description and Conservation. C. M. Finlayson, G. R. Milton, R. C. Prentice and N. C. Davidson. Dordrecht, Springer Netherlands: 1-13, 2016.


**Publications in conference proceedings**


8. **Hassaballah, K.**, Mohamed, Y., Uhlenbrook, S., and Omer, A.: Use of Hydrodynamic and morphologic models for the management of the seasonally flooded Mayas wetlands in the Dinder National Park-Sudan. International conference on Tekeze-Atbara Water Related Studies,25-25 March 2019, Khartoum, Sudan.
9. **Khalid Hassaballah**, Y.A. Mohamed and Stefan Uhlenbrook.: Maximizing the use of satellite data and hydrological modelling for streamflow prediction: Case

study of Dinder and Rahad rivers (tributaries of the Blue Nile/Sudan). Proceedings of Enhanced Flood Forecast and Early Warning System forum,16-17 August 2018, Nexus Hotel, Addis Ababa, Ethiopia.

10. **Khalid Hassaballah**, Y.A. Mohamed and Stefan Uhlenbrook.: Assessing the impact of land use and land cover change on streamflow response: case study of Dinder and Rahad, Ethiopia/Sudan. The 5th Nile Basin Development Forum 23rd to 25th October, 2017 Radisson Blu Hotel & Convention Centre, Kigali, Rwanda.
11. **Khalid Hassaballah**, Y.A. Mohamed and Stefan Uhlenbrook.: Floodplain Transformations and Mayas Evolution in Dinder National Park (DNP)/ Blue Nile/Abbay. International Conference on "Contemporary Evolution of African Floodplains and Deltas", Dar Es Salaam, Tanzania, 27 to 30 May 2014.
12. **Khalid Hassaballah**, Y.A. Mohamed and Stefan Uhlenbrook.: The long-term trends of the hydroclimatic variables in the Dinder and Rahad River Basin (Blue Nile, Abbay). Proceedings the New Nile Perspectives Conference, Khartoum, Sudan, 2013.

*Netherlands Research School for the*
*Socio-Economic and Natural Sciences of the Environment*

# D I P L O M A

## *for specialised PhD training*

The Netherlands research school for the
Socio-Economic and Natural Sciences of the Environment
(SENSE) declares that

## *Khalid Elnoor Ali Hassaballah*

born on 23 October 1975 in Sennar, Sudan

has successfully fulfilled all requirements of the
educational PhD programme of SENSE.

Delft, 9 November 2020

The Chairman of the SENSE board

Prof. dr. Martin Wassen

the SENSE Director of Education

Dr. Ad van Dommelen

*The SENSE Research School has been accredited by the Royal Netherlands Academy of Arts and Sciences (KNAW)*

K O N I N K L I J K E   N E D E R L A N D S E
A K A D E M I E   V A N   W E T E N S C H A P P E N

The SENSE Research School declares that **Khalid Elnoor Ali Hassaballah** has successfully fulfilled all requirements of the educational PhD programme of SENSE with a work load of 46.1 EC, including the following activities:

**SENSE PhD Courses**

- Environmental research in context (2013)
- Research in context activity: 'Co-organizing program and communication with participants for International conference on: New Nile Perspectives – Scientific Advances in the Eastern Nile Basin , 6-8 May 2013 – Khartoum, Sudan'
- Academic Writing Course for PhD Fellows (2018)

**Other PhD and Advanced MSc Courses**

- Training Course and knowledge sharing workshop on: Ecosystem in IWRM, Makerere University, Uganda (2012)
- Land Surface Process Modelling, Utrecht University (2013)
- Delft 3D Modelling summer course, IHE Delft (2013)
- Economic valuation tools for Wetlands ecosystems management, NBCBN, WaterCap, Kenya (2015)

**External training at a foreign research institute**

- Internship in the use of "WFlow" hydrologic model, Deltares, The Netherlands (2014)

**Selection of Oral Presentations**

- *Floodplain Transformations and Mayas Evolution in Dinder National Park (DNP)/ Blue Nile/Abbay*. International Conference on Contemporary Evolution of African Floodplains and Deltas, 27 to 30 May 2014, Dar Es Salaam. Tanzania
- *Assessing the impact of land use and land cover change on streamflow response: case study of Dinder and Rahad, Ethiopia/Sudan*, The 5th Nile Basin Development Forum, 23-25 October 2017, Kigali, Rwanda
- *Maximizing the use of satellite data and hydrological modelling for streamflow prediction: Case study of Dinder and Rahad rivers (tributaries of the Blue Nile/Sudan)*, Enhanced Flood Forecast and Early Warning System forum, 16-17 August 2018, Addis Ababa, Ethiopia
- *Use of Hydrodynamic and morphologic models for the management of the seasonally flooded Mayas wetlands in the Dinder National Park-Sudan*. International conference on Tekeze-Atbara Water Related Studies, 25 March 2019, Khartoum, Sudan

SENSE coordinator PhD education

Dr. ir. Peter Vermeulen